Petersen's Creative Customizing

Introduction

Welcome to the new Deluxe Series of books in the Petersen automotive library—new in both concept and level of presentation. Our on-going lineup of regular basic books is designed for the beginning auto enthusiast in the six sub-system areas of the modern car, one of which is on body repairs and painting. This first book in the Deluxe Series will, then, take up where its basic counterpart leaves off. It is aimed at the already experienced bodyman and painter who is well-versed in straight sheetmetal repairs and everyday painting, but who senses the urge to explore pent-up creativeness with shapes and colors of his own design.

To elevate the "bump shop" worker to the highs of what can be lucrative rewards in professional customizing, it seems apropos to briefly recap the field of automotive craftsmanship. It has its roots in turn-of-the-century carriage building, progressing to the motor vehicle as the engine began replacing the horse. In the introductory chapter we trace the bridge that early artisans crossed as they advanced through the eras of stately limousines, through the age of the low-slung, unmuffled raceabouts, and on up to the brink of The Depression when ostentation via the handcrafted automobile very nearly disappeared. There was a brief resurgence of the one-off-crafting of cars for the rich and the royal until World War II signaled the end.

Late in the Forties motorized flamboyancy returned, but in a new form—ordinary cars with just touches of personalization to represent automotive distinction. The term customizing replaced coachbuilding, and a new wave of artisans came to the fore—clothed in soiled jeans and T-shirts instead of the laundered smocks of their forebears. Rather than shape a virgin sheet of steel into a flowing fender to be fitted to a custom-bodied Duesenberg or Lincoln, these newcomers cut and hammered ordinary Fords and Chevrolets into sometimes illogical shapes, but more often into works of a new form of art.

Illustrations showing the development of customizing are included, not only to show that virtual masterpieces could be made from every man's car, but also to reveal the creativity that then existed in back-alley shops and ill-equipped garages. We hope some of these will stimulate today's customizers to new heights of craftsmanship.

More importantly, we present in how-to form customizing and painting techniques as they can be applied to the modern car, the van, and the pickup truck. Today, our transportation choices are more look-alike than ever before in history; hence the need for an increase of fresh talent in the customizing field. We hope this Deluxe Series effort will start you off on the wide-open road to the creation of automotive self-expression.

SPENCE MURRAY

COVER: Customizing as an art form is represented by three fine examples of creativity. In the foreground is Proto 1, an entirely handcrafted car. Center stage is a hatchback-styled Corvette, depicting today's popularity of add-on kits. At rear is a Datsun Z-car which combines mild bodywork and a minimum of custom painting, but which combine to result in outstanding self-expression. Chrome-styled logo by Alan Hashimoto. Cover design by Dick Fischer. Photography by Bob D'Olivo.

Library of Congress Catalog Card No. 78-50827

ISBN 0-8227-5026-0

CREATIVE CUSTOMIZING

Edited by Spence Murray and the staff of Specialty Publications Division.

Contents

CREATING SHAPES

CUSTOMIZING CHRONOLOGY IN COLOR

CREATIVE PAINTING IN COLOR

CREATING HUES

Coachbuilding to Customizing

Across 85 years of time

by Jay Storer PHOTOS BY PPC LIBRARY

Customizing, or modifying to suit personal tastes, has probably been around since the invention of the wheel, and certainly long before the invention of the automobile. From the beginning of horse-drawn vehicles, there were always carriage and wagon-building firms that in addition to their line of standard production bodies would build "specials" of a custom nature for wealthy clientele.

This was the basis for the custom automobile body business when the self-propelled vehicles came into their own in the late 1890's. Most of the pioneering inventors who designed early automobiles were mechanical engineers, men who had little time or inclination for the aesthetics of body building. So they would either design their gas buggy around a standard wooden carriage of the day, or turn that phase of their "industry" over to one of the custom carriage builders, who could build a body to the inventor's specifications and dimensions.

This pattern of farming out the bodywork on the early automobiles remained for several decades. Even when the industry developed to major status as the automobile became a necessity for every American, most manufacturers did not produce their own bodies. Even Henry Ford, an automobile magnate who liked to have every facet of his automobiles handled "in house," had a lot of his bodies produced by body building firms such as Murray Corp. of America. Murray was one of the largest body building companies, supplying bodies for such status cars as Lincoln, Packard and Reo as well as the more mundane Fords. Most of their business was in production bodies, but they also did custom work and had one of the first styling studios in Detroit. From the hundreds of coachbuilding and body building firms that were operating at one time, Murray is one of a handful that are still in business supplying components, although to a very different auto industry than when they started.

Most of the coachbuilding companies disappeared for the same reasons that most of the over-2000 car manufacturers disappeared. The Depression killed businesses—and tycoons—with a shotgun approach that hit the automobile industry hard. Only those manufacturers that were large enough and shrewd enough to produce cars at the lowest possible price were able to survive the dismal sales years of the Depression. Expensive luxury cars were difficult to sell when so many people were out of work and even the banks were in financial difficulty. A few manufacturers struggled on with the really expensive machines like the Duesenberg, supplying the members of the "carriage trade" that had so much money they could continue their extravagant ways even through the crisis.

Auto body building had its roots in the genuine "coach"-builders of the period; the transition from wagons to self-propelled vehicles was an easy one since most of the very early cars had wooden bodies. Later on, the use of wood was continued even after metal bodies became the rule, the wood being

1

used for structural support for the skin. Up until the development of quality stampings for large sheetmetal panels, all automobile bodies were built with a wooden framework that was covered with sheetmetal panels. These panels were actually nailed in place over the wood, with strips of semi-flexible metal moldings used to cover the seams. It was precisely because of this method of construction that automobile customizing as we know it was developed. Since a manufacturer or coachbuilder didn't have to invest hundreds of thousands of dollars in tooling up stamping dies for production body panels, which he would have to produce in large quantities to pay off, there was the freedom to vary designs easily. Many different body styles could be arranged and when a special order came in for custom body, it wasn't much more difficult to produce than a production body.

There have always been people "of means" who weren't satisfied to have a production body. They would have a special body built to their tastes and picked out of a set of designer's drawings, or they would have a production body modified or restyled. The extent of customizing varied from simple custom paint treatments to mild fender changes to a completely new one-of-a-kind body. The car manufacturer's catered to these custom body builders, or coachbuilders, by building a percentage of their production cars with only the main body in place; the radiator shell, hood, fenders and other accouterments to be added by the coachbuilder. Many of the luxury cars like Lincolns, Packards, Cadillacs and others were often delivered to various coachbuilders as a bare chassis, on which a completely custom body would be produced for a wealthy client. Some manufacturers, such as the most famous and exotic of American cars the Duesenberg, never produced a "production" body at all! They simply built a luxury chassis and powerful drivetrain as the basis for the coachbuilders to weave their sheetmetal fantasies on. The closest such manufacturers came to a "standard" body would have been when one of the coachbuilders built several similar customs.

DECLINE IN USE OF WOOD

The decline in the use of wood in automobile construction was hastened by the rise of the business to major industry status, requiring that more and more steel be used in construction and that body panels be stamped out by the thousands to meet demands. The automobile had become more than a plaything for the rich, it was in the 1920's a virtual necessity for every family. The first of the all-steel bodies was built for Dodge in 1921 by the Edward G. Budd Manufacturing

2

3

4

5

1. The epitome of the coachbuilding era were the custom-built bodies on the Duesenberg chassis, such as this Bohman & Schwartz convertible coupe for actress Mae West. Luxurious and fast, these cars were sold as chassis only, there were no "production" bodies.

2. Up until the late Twenties, most cars used a wooden framework for the body panels, and coachbuilt customs used wood right up to the Forties.

3. When cars had wood-framed bodies, building a custom body was no more work than a stocker; a new frame to the designer's specs was built and covered with metal, like this dual-cowl phaeton being restored by the artisans at California Metal Shaping.

4. While there were a few cars that were sold as chassis only, most of the coachbuilding business involved custom bodies built on standard luxury-car chassis. This carriage-style "panel brougham" by Derham is on a 1938 Packard chassis, built for famous opera singer Lily Pons.

5. By the Forties, most of the real coachbuilding had died. Custom bodywork stayed alive however, as there were still those people who wanted something different and sporty. Movie people seemed to gravitate to sporterized passenger cars like this 1940 Packard customized by Howard "Dutch" Darrin, one of the few designers from the early days who never stopped restyling.

Coachbuilding to Customizing

Co., who had been (and are still today) involved in producing railroad cars but produced auto bodies for Dodge and others up until the early 1930's. General Motors hung onto the heavy use of wood (up until the mid-'30's) longer than any other domestic manufacturer. Ford followed Chrysler's lead very quickly, and by the mid-1920's, most non-GM cars were built out of steel and lined with wood, rather than being built out of wood and skinned with sheetmetal. The wood in the bulk of cars was now being used merely to facilitate attaching hardware and upholstery. This decline in the use of wood for structural bodybuilding was part of the handwriting on the wall for the coachbuilders. The Depression just sounded the death knell for that fabulous period of new mores, new wealth and $15,000 custom-built automobile bodies. The imports such as Isotta-Fraschini, Maybach, Minerva, and Delage were among the first to fall, since most of them were sold only as a chassis for special-ordered coachwork. The production all-steel bodies were superior to the steel-over-wood designs in terms of strength, safety and freedom from rattles and squeaks, and the era of ostentatiousness that had created the custom-body market was dying.

Despite the end of major coachbuilding domestically (there are still custom firms in Europe), the spirit of individuality that had created that market never died, it just lessened in scale. It's part of the American spirit to want to be different, or at least look different, and just after WWII, it took hold of young Americans who had a few extra GI dollars and wanted to have their cars changed slightly to suit their own tastes. This was long before the period of arm's-length options lists that allowed Detroit's customers to virtually build their own car before buying it. Standard American sedans were fine for the bulk of the populace, who were hungry for any car right after the war (even facelifted '42 models), but for the returning GI and other young men who had been exposed during the war years to new, worldly influences, they were built like tulip bulbs. They yearned for the same qualities of form that have been the quest of annual facelift designers and ad copywriters since the beginnings of the automobile: longer, lower, smoother. None were well-heeled enough for the traditional work of Murphy, Derham, Brunn et al, but most could afford a few mild changes that at least marked their transportation as "youthful." At first, the mildest treatments consisted of merely removing some of the stock trim items and adding a few accessories purchased at a local Western Auto Store. As the Fifties progressed (?) with more and more chrome do-dads, this aspect of customizing became even more important and is considered de rigueur even today. Then there were always those who wanted something truly different, and there sprung up all across the country those specialized body shops that later were to be called "custom" shops. A few were staffed by some of the old-time craftsmen who had been out of such work since the end of the coachbuilding era. Other shops were started, or converted from standard body repair shops, by new young pairs of talented hands like Harry Westergard, Linc Paola or Jimmy Summers, faces on the scene in California, where automotive trends have a habit of starting then as now. There were still a few shops left in business, such as Coachcraft in Los Angeles, that

1

2

catered to the rich crowd, but they weren't involved in complete custom bodies anymore. Even coachbuilding had tightended its belt a few notches, to the point where most business was in modifying production bodies. Coachcraft turned out beautiful custom versions of Lincolns, Cadillacs, Chryslers and Packards, often for members of the Hollywood movie crowd.

A MAGAZINE BUILT A FIRE

With the coming of automotive enthusiast magazines at the beginning of the Fifties, the techniques and styles of customizing became not just California phenomena. All across the country, high school students neglected World History and Mathematics for such titles as Rod & Custom, Honk, Hot Rod and Motor Trend. Dozens more came and went during the Fifties, many of them in the 5"x8" format which seemed to fit just perfectly inside of a school book! Well up into the early 1960's, these magazines came relentlessly to the newsstands with new stories on how to "restyle from mild to wild." With their help and a local bodyman who understood more than fixing crunched fenders, a car could be built mild, in which case all of its trim would be removed and the holes filled in by welding and leading, the headlight rings would be "frenched" or molded into the fenders, taillights would be swapped for those of another make or tunneled into the body, the grille would be swapped for a chromed-tube item from the Western Auto Store or J.C. Whitney catalog, and this would be topped off with Moon discs or Oldsmobile Fiesta hubcaps, rear fender skirts and of course . . . pinstriping. If you had the talent or the money to hire such talent, you could go much wilder. The bulbous shapes and slab sides of those '49-'51 Fords, Mercurys and Chevys (by far the most popular makes to customize) really lent themselves to the difficult-but-rewarding major modifications such as top chopping, channeling, sectioning, "pancake" hoods, and a host of other treatments. Depending on the "competition" around your local drive-in hamburger stand, a customized car during the Fifties was a guarantee of social success, with the guys and the gals.

Certainly the customizing thing was started on the West Coast, but with the popularity of the enthusiast magazines and the growth of the car show, it wasn't long be-

1. Although they seldom built wholly new bodies for customers, Coachcraft was a true customizing shop. This incredible roadster built in the early Fifties on '40 Ford components shows flawless styling and workmanship.

2. One of the few coachbuilding firms to survive until today, Coachcraft of Los Angeles had their heyday in the Fifties. Their specialty was making sporty versions of passenger cars and adding the very popular continentals to new cars, as this '54 lineup shows.

3. Then there were the new faces that showed up doing customizing for the masses, masters like Joe Bailon, at left with his radical '41 Chevrolet. This crowded nostalgia scene took place at his busy Hayward (CA.) shop in 1951. At center is a chopped 1948 Futurama Oldsmobile with padded Hall top.

3

fore cars were being customized everywhere, and special custom shops opened up across the country. California of course had the majority of the famous names, like Joe Wilhelm of San Jose, Joe Bailon of Hayward, Gene Winfield of Modesto, and in L.A., the list went on and on: the Barris Brothers, Valley Custom (Neil Emory and Clayton Jensen), Dean Jeffries and many others. In other areas, those who made a name for themselves through exhibiting their work in magazines and car shows included Ray Wilson of Kent, Washington Kansas claimed the residences of Darryl Starbird of Wichita (also Dave Stuckey who worked for Darryl before going out on his own), and Bill Cushenberry (who moved his shop to Monterey, Calif. in 1962; Ray Farhner of Kansas City, Mo.; Dave Puhl in Chicago; the Alexander Brothers in Detroit; and Dick Korkes, whose busy custom shop was located in the unlikely spot of Whippany, New Jersey. Scattered far and wide once customizing become as American as rock-and-roll music, there were hundreds of much smaller but equally active "Anytown Custom Shops" who turned out the bulk of the minor customs that could be seen on any Friday or Saturday night at the local hamburger stand.

Of course, not all customs were built by shops. For every well-known show car built by one of the names, there were hundreds of mild customs that were driven daily by their owner/builders. Many were the dungareed youths who learned metalwork in the service or at a trade school just so they could get in on the fad, built themselves one nice rod or custom, and then never picked up hammer, torch or spraygun again. Modern restorers of special-interest cars would trade their subscriptions to Hemmings Motor News for a chance to go through one of the trash bins outside a high school auto shop class during the customizing period of '48-'65, crammed as these trash bins were with the perfect emblems and chrome trim from students cars.

Part of the spread of customizing as a social phenomenon was due to the way ordinary (non-enthusiasts like parents, police, etc.) people looked at them. Most of the true custom cars didn't have the hopped-up engines of the hot rod set, who gravitated to the earlier body styles like fenderless '32 roadsters and coupes (and hence stood out more to start with). Some of the early custom cars may have had a special intake manifold with dual carbs, some chrome accessory goodies, and a dual exhaust system, but the majority had stock engines and what equipment was added underhood was generally for show and "sound." Because of this deemphasis on engine perfor-

3

1

2

4

mance among the custom set, and because of the low ground clearance of the cars, most customs were driven in a careful manner. The hot rodders, on the other hand, for years had to live with an image only slightly cleaner than Hell's Angels types, because they were often exhibitionist about their car's performance, "peeling out" at every opportunity and even racing on the street. The cleaner image of the custom crowd meant toleration by the gendarmes and acceptability to parents, to say nothing of the "cool" status achieved with your peers, especially of the female variety.

GLORIOUS SHOWTIMES

Closely tied in with the history of customizing is the history of the car shows that featured such cars. There had been auto shows for many years, both in this country and in Europe, almost since the beginning of the automobile, but nothing like a custom show. In the immediate postwar years when car enthusiasm was high and the ex-GI's were building rods and customs, some took to an occasional display in the high school parking lot or near an auto store. In fact, the first custom shows were really the unofficial ones that you could find on a good Saturday night at the hamburger stand parking lot. Like anything that captures the imagination of a sizeable segment of the populace, it was inevitable that car shows would become more organized and commercial.

Customizer Gene Winfield remembers the first "real" show he had ever seen. Early in 1948 he had been (down from his Modesto shop) to Alex Xydias's famous SoCal Speed Shop in Burbank to gather some flathead goodies when someone said, "Why don't you stick around until Sunday, Gene? There's going to be a bunch of cars gathered at the unfinished Mercury plant in Pico Rivera. They're going to pick cars to be in a big show at the Armory." This gathering of SCTA lakes racers and street rods was the first unofficial car show we know of, and the show that it was the prelude to was the first "real" show of an organized nature. The show took place in January 1948 at the Armory in Exposition Park in Los Angeles, jointly sponsored and organized by the SCTA (Southern California Timing Association, a collection of car clubs who organized the dry lakes racing events) and Hollywood Publicity Associates. One of the partners in this latter sponsoring agency was Robert E. Petersen, founder of this publishing company. It was at this Armory show that he distributed the first copies of his new Hot Rod magazine, which later grew to become the world's largest automotive magazine. That first issue (January '48), which is a collector's item today, has a story in it about the plans for that Armory show, and a photo taken from the roof of one of the Mercury

5

6

1. One of the legends of those early customizing days was the late Sam Barris, brother of George who is still active in building custom movie cars.

2. A fine example of those traits is this chopped '41 Ford. Everything on the car is custom, from the Carson padded top to the '49 Olds grille bars and frenched headlights. It all came together under flawless lacquer. Only 47 ins. high, remarkable for the period, it was nicknamed by other enthusiasts the "four-foot custom."

3. Early metal master Harry Westergard, builder of this '39 Ford, was instrumental in teaching customizing to the Barris brothers. Now owned by photographer/custom fan Andy Southard Jr., this smoothie features chopped, removable hardtop and frenched-in seams everywhere. Note the filled decklid and sunken plate housing.

4. Custom-car enthusiasts hungered so for information and a look at new customs that they made car shows a successful business right from this very first show, the 1948 Motorama in Los Angeles. Note the crowd!

5. Besides the local drive-in and the car shows, interest in customs was kept high through the vicarious pleasures of Hot Rod, Rod & Custom and other enthusiast periodicals. More than one high school student of the Fifties found teachers didn't approve of them as in-class reading, but these publications helped set trends.

6. Coachbuilt, but in the custom-car era is this '49 Cadillac town car built by Maurice Schwartz of Pasadena. Smooth lines, almost total dechroming and even custom-made hubcaps.

PHOTOS COURTESY ANDY SOUTHARD, JR.

plant buildings showing some of the cars selected for the show.

From that beginning, the car show phenomenon spread rapidly. Actually, the Los Angeles area had the first show, but it took some time after that to catch up to the northern California region, where the shows really developed to their fullest. The first northern show, early in 1949, was a "hot rod exposition." This went so well that shortly thereafter the first big Oakland Roadster Show was held. This is still running every February, and has become the granddaddy of all car shows. Two years after the first Oakland show, some of the other northern California towns started their own shows, and thus began the first car show "circuit," with Modesto and Fresno being among the first of the other towns to participate. The rest of the country caught on very quickly, particularly in the Midwest and East, where there weren't any other car-oriented activities that could readily be enjoyed during the long winter months. Even in the West, car shows were more frequent in the winter months, and winter eventually became the show season all around the country.

As the 1950's went on, considerable space in enthusiast magazines was devoted to coverage of car shows from around the country. The snow-bound enthusiast in North Platte, Nebraska, could read about not just the famous Oakland Roadster Show, but shows in New York, Detroit, Boston, Chicago and Kansas City as well. The top customizers soon realized the commercial possibilities of the car show. The magazine publicity and public exposure that came with entering and winning shows meant more business for his shop, and perhaps the right to charge higher prices for his work. Although in the beginning there was little if any money involved in winning a show category (there were trophies of course), the competition in the custom car classes became stiff as each customizer tried to outdo the other. The custom fans who reached deep into their jeans to pay their way into these shows just ate it up, and when a magazine started a rumour that so-and-so was finishing up a radical to appear first at a certain show, that show was almost guaranteed a healthy gate.

Once the various car show promoters got their collective heads together, the scheduling of shows was worked out so that overlap was eliminated and thus some of the more specialized attractions could be featured at every show around the country. Cars of the famous names were always popular, as were any cars that had been given extensive coverage in an enthusiast magazine, or had been a magazine's project vehicle, such as Spence Murray's Rod & Custom Dream Truck.

Naturally, the display of a car became very important to show it in its best light for the judges. In the early shows, the cars were just parked, and each had to stand on its own merits, but it wasn't long before someone decided to build a fancy display for his car to attract more attention (the public often had a say in one of the awards, titled the "Peoples Choice" Trophy). It got so that show promoters had to create a special award category for the displays themselves. This created even more competition for the participants, who created more and more outlandish displays, making the judging of a car on its own merits for the class awards something very difficult to be objective about. Participants created spiralling displays in that same American spirit of competitiveness that made them show the cars in the first place. Some had their cars blocked up over a set of mirrors, so spectators and judges could see the detailing underneath, some had miniature waterfalls and other bucolic sets, and almost every car was surrounded with spun glass fibers known as "angel hair" that gave the impression the car was surrounded by fog or perched on a heavenly cloud.

1

HIGH-BUCKS DEMISE

Later in the 1950's, some of the Eastern show promoters began paying the top cars to appear in their shows. Carl Casper and Bob Larivee, two of the top promoters in the midwest, are generally credited with being the first to pay out for "circuit" cars that went to every show. At first, the promoters just paid "tow money" at so much a mile for the expense of travelling, but some top customizers became such drawing powers at the gate that they could demand even more compensation. In the later 1950's and early 1960's, the famous Oakland Roadster Show run by Al Slonaker used to give away a trip to Europe as part of the grand prize, but in 1963 a cash prize was substituted for the trip. In the furthest development of the commercial aspect of car shows, certain famous cars were "leased" by the promoter of a certain show circuit and the owner or builder paid a certain amount for each show, regardless of whether the car won a class or not. In fact, most of the cars in a show were local cars built by average guys, and they were so incensed at having to compete with professionally-built machines imported from "The Coast", that promoters started showing these top cars strictly as draw cards. In other words, these cars were shown, but not entered in competition for any of the class trophies.

While this satisfied both the local competitors and the crowds who came to see the famous cars, it also contributed to the development of professional show cars, and in a way to the demise of customizing in general. When the professionalism really took hold, winning a show award or getting top billing degenerated into a money game. It was no longer a question of who could build the nicest and most original car, but who had the most money to spend to achieve results. If you had enough money, you could hire a top customizer to build your car for you, and also have the most elaborate display. Someone could, and did, come to George Barris and say, "I want to win the Oakland Roadster Show, and I don't care what it costs."

Two other factors that contributed to this demise of the street-driven custom car were upholstery and chrome plating. Early in the custom period, a simple tuck-and-roll interior was all that was ever needed, and chrome was thought to be something you took off of a car. But somewhere along the line, show competitors figured that if some is good, more is better, and they went wild with Naugahyde and chrome. Pleats and rolls showed up on panels that snapped into the wheelwells of cars and under the hood, even over the radiator. Trunks were not just carpeted but fully upholstered and outfited with a display of chromed tools and gas and water cans. If chromed road flares had been available then, they would have appeared in every show car's trunk. Early customs had only a few chromed accessories to decorate the engine compartment, but this degenerated to the ridiculous later on, until everything under the hood received the shiny treatment. At the furthest extreme, there were those show cars that had completly chromed undercarriages, and some sported hoods and fenders that were entirely chromed on the engine compartment side. Obviously, this kind of show-boating made the cars less practical for true street use and thus developed a large group of cars that were "for show only." Once a car fell into this category, then the sky was the limit as to what modifications could be done, and "everything" was tried. Metalwork and paint alone were enough to make a

1. Deep maroon and deeply channeled, the '40 Mercury convertible of Jimmy Summers remains one of the classic early customs. Summers, along with Westergard and Paola practiced the customizing arts in the mid-Forties as we know it now, cutting, bending, welding and reshaping production cars to take over where the coachbuilders left off.

2. A rare look inside a Fifties-era custom shop finds car owner (center) Jack Stewart conferring with Valley Custom Shop partners (and brothers-in-law) Neil Emory (left) and Clayton Jensen. Subject at hand was the yet-to-be Polynesian, a sectioned 1950 Oldsmobile hardtop (shown in color on page 83) that became one of the most famous customs of the early Fifties.

3. While Clay works on the Polynesian (foreground), Neil finishes welding a chopped windshield post on his nephew's 1940 Mercury convertible, while a helper dings out a fender and another contemplates the engine compartment of a Model A roadster.

4. Outside, amidst enough neat old Fords to set a modern enthusiast to foaming at his swap-meet-parched lips, sits (low) the nearly finished 1950 Ford of Ron Dunn, another famous section job. Although not chopped, all glass was removed to insure a perfect lacquer paint job.

5. Sectioning was one of the most difficult custom jobs, but applied to the '49-'50 Ford gave dramatically pleasing results. This is Dunn's car the first time. Like many customs, it received a later facelift. A color shot of the car is on page 83.

2

3

4

5

neat car in the early days, but somewhere along the line someone put a GMC supercharger on top of the all-chromed engine in his custom and the end was near. Since the car didn't have to be driven, many show cars were built that could not even run, and there were more than a few participants who sold the internal parts of their engines because they didn't show, presumably to pay their astronomical plating bills for the outside of the engine. If one supercharger was good, two would be even wilder; if your competition had dual carbs, why you needed to put eight on your car. A few shows, like the respected Oakland Roadster Show, kept a certain sanity to the affair by requiring that cars entered in competition for any of the awards be capable of being driven into the auditorium, but this attitude on the part of some promoters wasn't enough to stem the tide of trailered-only cars.

The final death blow to customizing was the advent of the Detroit "muscle car" in the mid-1960's. Beginning with the 1964 Pontiac GTO and continuing almost relentlessly, Detroit kept inventing new cars and new marketing strategies to appeal to the younger car-buying public. Drag racing and performance had after a long struggle shed their "crazy speed-punks" public image, and flourished dramatically once they achieved respectability as a hobby/sport. Everyone was reading Hot Rod Magazine and trying to make their cars perform better. Popular songs became hits when they were about performance: Little GTO, My 409, Spring Little Cobra, Tach It Up, and others were on the lips of every high-school-age male. The backroom full of marketing boys in Detroit finally took notice of this trend and packaged some of their regular cars to appeal to this market. The GTO was basically just a "treatment" given to a standard Pontiac Tempest. What the auto moguls found out was that "performance sells," and these were the bywords they taped on their stylists', engineers' and marketing specialists' walls. The engineers began building performance packages for every car line, with special engines and 4-speed manual transmissions, heavy-duty "handling" packages for the suspension, and the stylists in turn wrapped these cars up with scoops, stripes and other distinctive external identification. The factories were getting heavily involved in drag racing, and the enthusiast magazines reflected this change in attitudes. The space many of them had devoted to results of car shows around the country was being replaced by coverage of the various drag events, and features on customized cars were edged out by features on race cars or road tests of the latest Detroit entry in the horsepower race. Even some of the top name customizers found a complete reversal of their business. Instead of building custom cars for individuals, they were doing contract work for the auto makers, building vehicles for the factories to bring to car shows to stimulate young car buyers. The Ford Custom Caravan of '63-'65 is a perfect example.

NEW DIRECTIONS

There's no sharp dividing line where customizing lost public attention in favor of performance, but it would be safe to say that by '65-'66, the emphasis was off of custom body modifications. Detroit auto manufacturers, the same group that contributed to this decline, was also responsible for it's rebirth almost ten years later. Customizing is definitely back to stay, but it has taken new directions. By about '72 or '73, government restrictions on exhaust emissions and

1

2

insurance company pressures to lower horsepower had all but emasculated the Detroit muscle cars. It was in this period that the enthusiasts started bemoaning the lack of excitement in factory-built vehicles and the realization set in that if they wanted a performance vehicle they would have to build one themselves. Unfortunately, State and Federal emissions laws have had something to say about this trend too, and most performance modifications were illegal, even when performed by the owner of the vehicle or his local garage. Car enthusiasts are a hard bunch to keep still, and this kind of restriction in one area meant that if they were to continue to have personalized transportation they would have to go back to customizing. Body changes and custom painting were back in full force.

In fact, due in great part to the explosion of interest in pickups, vans and RV's, the custom paint field may be even bigger today than ever. The vans and trucks appealed to the former performance enthusiasts in a new way. Instead of playing drag hero at the stoplight, they played off-road racer on camping trips to the boondocks. The slow-performing vans appealed to a new group, the street cruiser types, and both the trucks and vans offered huge expanses of plain sheetmetal that cried out to be personalized. The steady growth of the mag wheel business and the increasingly large tires offered by aftermarket tire companies meant that bodywork of some had to be performed to fit these large "feet."

Thus customizing was brought back in new ways, and is going strong today, which is what this book is all about. Once again, high school auto shop trash bins are full of auto nameplates and stock headlight buckets, and trade school courses in bodywork are booked up regularly. One of the main new directions that customizing has taken recently is the "road-racer" look. Enthusiasts want their cars to look lower, sleeker and more aerodynamic, with touches of the Trans-Am or IMSA sedan racers. Low-profile fat tires and wide wheels, enlarged fenderwells and even complete custom fenders, spoilers, air dams, hood scoops and the new rectangular headlights have become the new stock-in-trade of customizers. Just as in the custom heydey of the 1950's, the radical jobs like chopping and sectioning are still being done, especially on trucks and vans, but the majority of enthusiasts set their sights a little lower, and content themselves with trick paint and racy fender treatments.

Another reason for the rebirth of customizing today is the current interest in nostalgia for the 1950's. Ever since the movie American Graffiti captured everyone's imagination, more than a few memories were jogged as to what it was like in the heyday of the original cruisers. Consequently, many older customs have been brought out of mothballs and restored to their previous splendor, to the total fascination of those who were too young to have appreciated them back then. This nostalgia phase has gone so far that many cars are being customized today exactly the way they were done back in the Fifties. It's a fair bet to say that today there are more chopped '49-'51 Mercurys, for instance, than there ever were back in the early custom days. Most are equipped with tape players crammed with oldies-but-goodies tapes that further bring tears to the eyes of the nostalgic. A lot of innovative work is being done today by truly creative individuals, as well as a host of "instant" customizing kits and parts available, and whatever you're driving, mini-car or motorhome, we've got some ideas for you within these pages.

3

4

5

1. It all started with just a little angel hair. The car shows, which had done so much to further the growth of customizing, unwittingly did much to bring about its downfall.

2. Both the cars and the displays eventually got out of hand, as these two examples indicate. The oriental phone booth is covered with Formica!

3. The Detroit auto manufacturers got into the customizing act quite heavily in the Sixties. Besides their normal "factory show cars" built in their own styling studios, automakers had special customs built by the top name customizers for the show circuit. Because of union rules and red tape, it was much cheaper for the factories to go to outside customizers.

4. In the early Sixties, one of the finest examples of factory car show participation was the Ford Custom Car Caravan. A veritable fleet of customs, like this Winfield-styled Falcon, were built by top names for this series of Ford displays. They put the Ford name in front of many thousands of show-goers, and these cars were frequent magazine features.

5. With the beginning of the "smog era" in the Seventies, interest on the street returned to customizing, particularly with the still-growing camps of truck and van enthusiasts. Customizing today is divided between the colorful truck-and-van movement and the trend to restyle passenger cars into sporty "cafe racers" with bold, graphic custom paint jobs and add-on fiberglass body panels.

In a Manner of Speaking

Useful words and phrases in tin-bending language

by **Spence Murray** PHOTOS COURTESY PPC LIBRARY

A rare combination of most of the major customizing alterations described here. This 1939 Ford has been chopped, channeled, frenched, decked, filled and, judging from its low stance, the frame has been stepped in the rear. The channel job also required that the hood be sectioned.

CUSTOMIZING began acquiring a jargon all its own in the early stages of its development. Of course, there was already much specialized terminology that had been handed down from the more formal coachbuilding trade. But we're speaking here of useful slang terms developed to describe a difficult metal-working operation in a single word or at least a brief phrase. By the dictionary's definition, for example, "channeling" is defined as "to form, cut or wear a channel in." In custom-car use, however, channeling is the act of cutting a car body's floor loose, lowering the body down over it, then rewelding the flooring back in.

Most of the customizing terms given in this selected glossary are familiar to car enthusiasts, but some are commonly misapplied while a few may have been all but forgotten. To help keep the customizer's vocabulary alive, then, here are some definitions that have come down from as long ago as the late Forties.

1. Swallowtail door (with door reversed)/1938 Darrin Packard for Clark Gable

2. Swallowtailed 1941 Darrin Packard/standard door

SWALLOWTAIL DOORS are those which have had their upper edge (on a roadster or convertible) cut down, usually on a taper slanting rearward. Usually the quarter panel's upper edge was left at stock height, but on some versions the slanting of the door line was carried into the quarter panel before angling sharply upward again. The idea was to create a more sporting or "open" look to a car. Credit for the innovation goes to Howard "Dutch" Darrin, an auto stylist and coachbuilder of both pre- and post-war periods, and who produced many handcrafted one-offs and limited-edition specials for his wealthy clientele. Although the design has been carried over to many makes of cars, it was Dutch's famed convertible Packard Darrins that sported the treatment best.

1. Nosed one-piece hood/1958 Chevrolet

NOSING is the act of filling hood ornamentation holes and/or the seam between the hood halves. In the late Thirties and throughout the Forties hoods had become larger than then-extant factory presses could handle as a single stamping. Most therefore were assembled from two mirror-image halves butting along a central line that ran from the top of the cowl to the center of the grille. Most such hoods were riveted along the mating flanges, often with a molding that was T-shaped in cross section sandwiched between the hood halves. On many cars, this molding extended to the forward crown of the hood and ended with an ornament. This left the seam between the hood halves exposed from the ornament down to the top of the grille, though it was partially covered by a medalion or nameplate.

Although numberless customizers eliminated the hood seam by removing the central molding and ornament, then brazing or welding then leading the seam over, it was often inexpertly done with a wavy hood the final result. An easier way to achieve the "smooth" look was to fill the seam on the front of the hood only, from the top of the grille to the hood ornament. The distance was short, and because of the crowned shape of the hood front, there was seldom warpage. Too, most cars, even after hoods became single-piece stampings, carried an emblem which, when removed, exposed fastening holes. Whether the hood is one-piece or two, the filling in of holes and seam (if any) is termed nosing.

1. Open-car chop job/windshield frame only/1936 Ford roadster

2. Post misalignment during chopping/1977 Chevrolet pickup

3. Severely chopped hardtop/1957 Ford

4. Van chop job/1971 Chevrolet

CHOPPING is undoubtedly the most used term in the customizer's vernacular; but it is sometimes used to infer the cutting-into of any body panel. Chopping, in its common usage, is the lowering of a car's roofline by cutting the desired amount from the roof supporting structures; the windshield, door and window posts, as well as the upper quarter panels. Chopping an early-model car is a relatively straightforward operation, since until about 1932 most cars were very nearly vertical in general configuration and no difficult metalworking was required to lower the roof by 3, 4 ins. or more. But as cars developed to include a semblence of streamlining, with rakes to both the windshield and the top's rear, the job became increasingly difficult.

As a turret top is lowered, after removal of the desired dimension from the supporting structures, the upper post stubs misalign with the bottom ones in direct proportion to the amount of the chop. Compensation can be made by reworking the upper and lower post stubs to a more rakish angle, but with the severe angles that came along about 1941, chopping almost automatically meant the stretching of the roof.

On a late-model car, truck or van, the customizer will have to slice the top into two sections at a point usually near the rear of the front doors. Now the post stubs can be aligned but a strip of metal will have to be added across the roof's midsection. On cars with severe tumblehome (the inward slant of the top above the beltline), the roof may have to be quartered. Next to sectioning, this is perhaps the most difficult job the customizer will face. The unsupported roof is greatly affected by the heat of welding/brazing and warpage will be severe. Tack-welding the abutting panels at widely spaced intervals, filling in the gaps after the welds have cooled, then reheating and hammer-welding the seams, is the proper way of executing the job. Because of the great amount of hammer-welding needed, the term "hammered" has become synonomous with chopping. Hacking is yet another alternate term.

Novice customizers envisioning a chopped top most likely see only the end product in their mind's eye. But part and parcel of the operation is the cutting down of the windshield itself, and the side glass and backlight if the body style is other than a roadster or convertible. Wind wings and frames, on cars so equipped, (and which are usually potmetal and thus difficult to reweld), and reducing the height of the inner window frames or garnish rails, is as difficult and time consuming as cutting just the exterior body portions.

Curved windshields, to say nothing of curved side glass and backlights, are very difficult to cut. Windshields, by law, are a lamination of two pieces of glass sandwiching a sheet of clear plastic (to prevent flying glass in the event of an accident). Side and rear glass is molded to shape with built-in stresses that cannot be upset.

A glass specialist will have to be saught out to cut down a glass; not many will accept the job and those that do work only at the owner's risk. To cut a windshield, some specialists will score both sides of the glass, complete the breaks by cracking, then heat the windshield to let the sandwiched plastic expand sufficiently to allow a razor blade to be inserted in the cut line to slice the plastic. Because the amount of a chop job can only be taken from either the top or bottom of the glass (or most usually both, which doubles the risk of breakage), it is obvious that at least some reworking of the windshield header and the cowl flanges will be needed.

Some glass firms have been successful in sand blasting the unwanted strip of glass away, and molded side and rear glass is more safely done in this fashion. Before this technique was developed, however, some customizers resorted to dropping the side and rear glasses into reformed channels built lower down inside the body. This was especially the case in the Fifties with the advent of wraparound rear glass which could not be chopped.

The customizer, before tackling a chop job, should first consult with a glass specialist; he may dictate the physical limit of the chopping, as determined by the year, make and model of the car.

In a Manner of Speaking

1. Butterfly doors/General Motors Firebird

BUTTERFLY DOORS are usually, but not always, hinged along their top edge, instead of at the front or rear as in conventional cars. The idea originally was to create as large a door opening as possible on a very low car, and to lessen to a degree the amount of stooping needed to enter the passenger compartment. The most prominent and best-remembered production cars to use butterfly doors were the famous 300 SL Mercedes coupes, but the technique has been used with various degrees of success on many custom cars, one-offs and show cars. The innovation, while dormant in customizing circles for a few years, is currently finding favor with van owners who are butterflying not only the driver's and passenger's doors, but the side and rear cargo doors as well. The idea here, aside from the unique treatment, is to provide a sort-of sunshade, while the van is parked, with the door(s) in the raised position.

The short-lived Bricklin sports cars, as well as the famous Ford GT race cars, also used butterfly doors.

The term is also correctly applied to any car door which, though hinged conventionally, wraps up into the roof. The Corvette Sting Ray coupes, starting in 1963, have butterfly doors, as do many luxury limousines, ostensibly to ease exit and entrance. Some customizers have used the idea on cars that have been chopped, channeled and/or sectioned. It is a difficult project in these cases, however, since the supporting side roof structure and the tops of the door frames must be completely reworked and some provision found to divert water since conventional drip rails cannot be used.

1. Lowering at its extreme/1957 Ford hardtop

LOWERING is synonomous with dropping, and is in such widespread usage that an involved explanation is not called for. Simply put, lowering means the reduction of a car's chassis ground clearance by any expedient means. It does not directly involve metalworking by the customizer, but some may be required as a result of lowering (see Radiusing). The way in which a car may be best and most safely lowered depends upon the type of suspension employed; coil springs, leaf springs or torsion bars; and whether one is faced with independent suspension or a solid axle. Lowering blocks, longer spring shackles, reversed spring eyes, dearched springs, offset spindles, moving a beam axle from beneath its springs to the top, the relocating of a spring perch or cup—all of these methods and more are commonly called for. Though the experienced mechanic can plan and carry out a lowering job, the procedures are best left to a competent wheel alignment and/or frame specialist who has the knowledge to do the job right, the equipment necessary and, most importantly, the devices and tools to check and reset proper wheel alignment once the lowering has been accomplished.

Lowering or dropping of a radical nature may call for special frame work. See C-ing, Z-ing and Stepping.

1. Frenched lights/filled doorhandle and other holes/peaked hood/1949 Mercury convertible

FILLING, FRENCHING and PEAKING, while not synonymous, are at least closely related. Filling, as the word suggests, means the filling-in of body holes or the seams between abutting panels, especially on the two-piece hoods of the Forties and Fifities. Lead was, naturally, the filling medium in the early days of the hobby, but certain plastics made their advent in the early Sixties and now, of course, Bondo is the most widely used filling material. Leading and dechroming were alternate words.

Frenching, for an unexplained reason, was never applied to the act oı filling the seam of a two-piece hood. But it is properly used to describe the filling-in of all other seams; as, between rear fenders and quarter panels, grille shell and front fenders, and so forth. The term also applies to the joining of a trim piece, as a headlight bezel or taillight rim, to the adjoining sheetmetal body panel. To effect the latter, the (usually) plated trim piece must be electrically dechromed (or by judicious use of a body grinder) before welding or brazing it in place. The welds or any gaps between them are then leaded or otherwise covered, filed smooth, then the area is painted body color. The result is a continuous sweep of sheetmetal right to the edge of the headlight or taillight lense. Sometimes, especially with headlight frenching, the lamp unit's retaining and adjusting screws have to be reversed and installed from inside the fender since the bulb was no longer accessible in the conventional manner.

Peaking is the term used when, instead of a two-piece hood seam being filled in the conventional manner (on a relatively flat plane), the lead (or other filler material) was built up into a ridge marking the former seam.

1. Decked lid/1958 Chevrolet

2. Deck job/1956 Corvette

DECKING is the removal of ornamentation from a trunk or decklid. Car manufacturers have always considered the decklid an ideal place to mount their corporate symbols, nameplates, other brazen ornamentation, and license bracketry, sometimes to overshadow and sometimes to highlight the trunk handle and the lock which secures it. With the "smooth" look the intent of most customizers, all the rear-end geegaws and baubles must come off—which means, of course, that all manner of holes, slits and other openings through which the various fasteners passed are now exposed. The filling of such holes is termed decking.

The problems of a proper decking job don't end with filling the openings by welding or brazing. Some new method must be found for shutting and latching the decklid and a new location found for the license and its mandatory light. It is often the practice, after removing the trunk handle but retaining its latch assembly, to rig a solenoid to unlatch the lid electrically from a dash-mounted button, but hydraulic cylinders or electric screwjacks (usually from convertible top mechanisms) are sometimes employed to raise and lower the lid.

The license, if it had been deck-mounted originally, can take a variety of locations. Most often it is mounted in the center of the bumper and framed with a U-shaped guard containing a small light. But customizers have also used a chromed frame to house the license and fasten it directly to the decklid—which sometimes can serve to hide some or all of the ornamentation holes. On cars of the Forties, it used to be the practice to cut a rectangular opening in the deck in which to inset the license, then cover it (usually illegally) with a pane of glass.

1. Paneled rear quarter window/1940 Ford coupe

PANELING loosely describes the addition of a panel, usually of metal but sometimes of wood. The panel may be a substitute for one that has been discarded due to rust or extensive damage, or to create an entirely new shape. Most often, though, the term is applied to older cars which had a fabric insert in the roof but which has had a sheetmetal insert added.

Paneling is also used to describe the filling-in of an opening, especially a quarter window. The modification gives the car a more formal look since the result emulates the rear roof grace of an expensive luxury car. The word is synonomous with blanking.

1. Bobbed fenders/channeled and sectioned '40 Ford

BOBBING usually infers a size-reducing operation of a body panel, but most often applies to fenders. It means the shortening or trimming away of a fender's extremities. The job is frequently done on a radically lowered car to regain adequate fender-to-ground clearance, but can also be applied to early cars where, for appearance sake, the running boards have been removed. The ends of the fenders that formerly mated with the running boards, when they are reworked to a more pleasing or "finished" appearance, are termed bobbed fenders.

In a Manner of Speaking

1. Tunneled (and frenched) headlights/ 1957 T-Bird

TUNNELING, as the word infers, is the recessing of an object as a head or taillight into the surrounding sheetmetal. Though lights are the most commonly tunneled, the term can also be applied to grilles, antenna mounts—in short, any body accessory. Usually applied to lights in conjunction with frenching, the practice reached ridiculous proportions when some customizers tunneled their headlights up to 12 ins. into the fenders.

1. Stretched rear fenders/1946 Ford convertible

STRETCHING, in its broadest definition, infers the lengthening or widening of virtually any body panel by adding to its expanse for whatever reason, or by hammering it to a new, more highly crowned shape. The stretching of car's roof, for example, is often necessary when chopping a top since, as the top is brought downward, it must be lengthened (and also sometimes widened) in order for the door and window posts to retain their alignment.

The most common usage of the term stretching, however, is in reference to fenders which, for stylistic reasons, have been lengthened to enhance the profile of a custom car.

1. Channeled '40 Mercury stock fender height/sectioned hood

CHANNELING is usually described in abbreviated form, but there is more work to it than meets the eye. Of course, the complexity of the job varies from one make of car to another, just as it does between years of the same make. Also, there can often be a choice of ways to perform the operation, again depending upon the year and make.

Channeling will reduce the height of a car when lowering of the suspension doesn't satisfy the owner's idea of low. Chopping the top isn't really an alternative, nor is sectioning, since these jobs change the car's basic proportions. Channeling will reduce interior room, but from the viewer's standpoint the car will look about as it did originally except for ground clearance.

The job involves cutting the floor loose from the body sides, dropping the body the desired amount, then rewelding the floor back in. Often, the body mounting brackets on the frame will have to be repositioned and/or others added to strengthen the lower body sides. New sections of flooring will have to be added inside the doors, both for strength and to conceal the frame which will be exposed at the doors' sill line, and again for the same reasons inside the lower edge of the trunk opening.

On early cars, the firewall will have to be notched to allow for the frame side rails or, on later cars, it will have to be sectioned an amount equal to the body drop. This is done through a vertical portion of the firewall to leave the angled toeboard intact. The portion of the firewall below the sectioning line must be cut loose from the cowl, then rewelded after the body has been dropped.

There will be alignment discrepancy when the fenders and hood

2. Early-model coupe/body sides dropped over frame

are remounted. On a nonfendered, early-bodied street rod the front fenders are usually left off or the wheels partially circled with cycle-type fenders bracketed to the brake backing plates. But on a later car, the fenders are usually left in their original relationship to the frame, as is the grille, and the depth of the channeling job removed from the vertical sides of the hood!

Rear fenders, when retained, can drop down with the body but this leaves inadequate rear tire clearance. More often, the fenders are left at their original height and the body allowed to drop between them. Good wheel clearance will then mean reshaping the inner wheel wells.

Often, especially on cars of the late Forties, the customizer channels the body, leaving all four fenders at their original height. Then, an amount equal to the depth of the channel job is trimmed off the lower body sides. This gives the same visual effect as sectioning, but on some models it is easier to do.

Few modern cars can be channeled due to the widespread use of unit-body construction, but late-model pickup trucks can be and with very satisfying results.

1. Radical rear end lowering through stepping/1956 Chevrolet

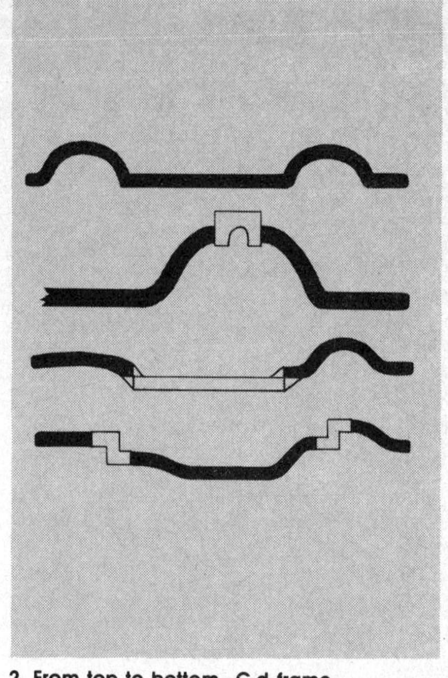

2. From top to bottom—C-d frame rear/Stepped frame/Z-d frame

C-ING, STEPPING and Z-ING are oft-confused terms, misunderstood, or mistakenly considered synonomous. They will be defined one at a time under this common entry since they all have to do with a car's frame, each operation involved with chassis lowering.

Because a car's frame is its backbone and provides a base platform for the body as well as the foundation for the running gear, it must be treated with respect and kept in perfect alignment. It is important that any modifications be done properly. While the customizer should be familiar with the steps outlined here, it is recommended that such work be actually performed by a chassis specialist who has the knowledge, tools and equipment to correctly prepare an altered chassis.

C-ing (also called arching) came into popularity in the early-customizing days when radical lowering through reworked springs brought the car's frame into dangerous proximity with the rear axle. The situation resulted in easy "bottoming" when the car passed over a rut since the axle-travel limit was virtually nil. The situation could be partly overcome by removal of the rubber snubbers designed to soften the bottoming effect. This gave about 2 ins. added axle travel, but was insufficient for over-zealous lowering fans.

If the portion of the frame side rails that contacted the axle could be eliminated, an unsung customizer reasoned, the bottoming problem could be overcome. And it was; by welding steel plates to the frame directly over the axle, boxing them in with more steel, then cutting a C-shaped portion away (see sketch). More axle travel was thus provided, and the customizer could lower away to his heart's content—so low in many cases that casters were sometimes installed under the rear bumper to help navigate steep driveways. But the pioneering C-er soon found a new problem; the frame of the car was so low in relation to the driveline that the driveshaft housing rubbed against the under-floor, necessitating the enlargement of the driveshaft tunnel toward the rear of the body, reworking the underbody cross supports, and modifying the rear seat. Early customizers would stop at nothing, though, and jobs of this type became quite commonplace.

While C-ing is generally limited to the rear of the car, Stepping can be done at either or both ends simultaneously. The job is done—and was almost always applied to the early Fords with their unique cross-springing—by severing the frame just ahead of the firewall, and again just behind the rear seat area. The central portion of the frame, carrying the bulk of the body, is then dropped a distance equal to the frame's depth, or some 6 or 7 ins. Fish places were then welded in place (see sketch) to hold the top of the central portion of the frame in alignment with the bottoms of the front and rear frame extensions. The car was thus dropped by an amount equal to the frame's depth, yet the front and rear suspension systems remain unaltered and ride and handling are not affected. Much metalwork must ensue, however, for the trunk flooring is raised inside the body, and front fender panelling must also be modified to make up for the depth of the body drop. So, too, must the driveshaft tunnel in the floor be reworked, for the shaft now passes through the body at a higher elevation than it did originally.

Z-ing found early popularity with the modified race-car crowd, using as they did primarily Ford products with the buggy springs. But customizers were quick to spot the idea and adapt it to street-driven cars most, but not all of which, bore the Ford nameplate.

In Z-ing a frame, the side rails are cut at a point just behind the front crossmember, and just ahead of the rear one. The frame's extremities, carrying the Ford spring-mounting crossmembers, are raised to any desired height (or, to state it another way, the central frame portion is dropped down). Heavy-gauge steel fishplates (usually in the form of the letter Z), are then added (see sketch) to hold all the frame sections in proper alignment. This technique, depending upon the degree to which it is carried out, can literally put the car's main frame portion right on the ground. Little bodywork is required up front after Z-ing, since fenderless roadsters were usually the recipients of the job with no interfering sheetmetal, but Z-ing the rear end means the elimination, or severe revamping, of the trunk floor.

None of these terms can be properly applied to work on a modern car since most are of frameless, unit-construction. But the ambitious customizer of today could consider the operations on some late vans and full-size pickups.

In a Manner of Speaking

1. Fender flared by spreading metal outward/adding conduit lip

FLARING, as the word implies and as the dictionary tells us, is "to open or spread outward." Although the term is most commonly used in reference to wheel openings that have been stretched outward (or where an add-on flare of metal or fiberglass has been attached), it is also correct to apply it to any body panel with one or more edges that have been spread outward by the customizer.

1. Pancaked hood/1956 experimental Oldsmobile F-88

PANCAKED HOOD is one which, instead of being deeply crowned, has been reshaped—or fabricated from sheetmetal—to a more nearly flat shape. It is often necessary to pancake the hood of a sectioned or channeled car since the amount of the body drop must be compensated for along the near-vertical sides of the hood. Performing this operation to a hood, especially on cars of the late Thirties and Forties, was so difficult that many customizers simply discarded the original hood and started from scratch with a sheetmetal panel cut and shaped to fit the engine compartment.

Many experimental and one-off show cars, when built very low in a sporty motif, carry pancaked hoods simply because the engine compartment opening is on the same plane as the front fender tops.

1. Well-executed sectioning/1949 Ford coupe

2. Rare early-model sectioning/1939 Ford convertible

SECTIONING is the most difficult of all customizing operations, and is therefore the least seen. Although it can be performed on any car, early or late, and on vans and pickups as well, it seems to look "right" only on certain models; the slab-sided Fords of '49-'51 were (at the time) the most popular, and easiest, cars for sectioning.

The term refers to the removal of a horizontal strip of metal from the car's body below the beltline. When the upper and lower halves of the body are reunited, the car takes on a lean, lithe look. Sectioning is sometimes accompanied by top chopping and/or channeling, but because each operation serves to severely reduce interior headroom, only a very few customizers have performed the three jobs on a single car.

Sectioning can be visualized by considering a football from which a strip of, say, 2 ins. is to be removed. If the strip is taken from near one end, it's obvious the upper part will no longer mate with the lower one. But if the parallel cuts are made 1 in. above and 1 in. below the center line, the two halves will butt together.

Obviously, a car must be sectioned at its widest part, with half the width of the sectioning dimension being taken from above the wide point, and the other half from below it. On the '49 Ford, body configuration allowed sectioning to be done on almost a true horizontal plane. On other cars, though, the cut lines had to be made high up on one panel, lower down on another, and sometimes at angles.

Not just the car's exterior skin is sectioned, but all the inner body structure, supports, brackets, and so forth, must be similarly cut then rewelded. Such cuts may not always be on the same plane as the cut through the skin, so the customizer must be well versed in all phases of metalwork to tackle the project. One way to "cheat" on a sectioning job is to drill out the spotwelds joining a door skin to the inner structure, section only the structure, respotweld the skin back in place and trim its excess from the bottom. This eliminates having to cut then hammer-weld at least a part of the car back together, although there are not many instances where this is possible.

Sectioning can lead to such frustration that more than one car during the process had to be junked when the customizer gave up. The end result of properly executed sectioning, though, is a custom guaranteed to stop other professional customizers dead in their tracks.

1. Radiused rear wheel cutouts/1960 Ford

RADIUSING can either be a fully practical or merely an esthetic operation, and is often, but not always, a needed operation after severe chassis lowering. It involves the enlarging or opening up of the wheelwells. On radically lowered cars, radiusing the front fenders may be needed in order to retain the full steering travel. But it is just as applicable where, for example, a car with squarish wheel cutouts is modified to more nearly round ones. On some cars, customizers altered the rear wheel cutouts in order to match the style used in front if the designers had not done so originally.

Radiusing is a fairly straightforward job for the customizer since usually all that is required is to cut away the unwanted or offending part of the fender with tinsnips or air chisel. However, necessary fender edge strength must be retained by reforming a flange or "curling under" of the edge around the opening. Modern customizers, when radiusing wheel openings to accommodate oversize wheels and tires, often form a new radius with electrical conduit, hand-bent around the tire, which is then welded/brazed in place and the gaps, where they exist, filled with sheetmetal.

1. Eyebrowed front fender/1956 Chevrolet experimental Impala

EYEBROWS are extensions usually added over wheel openings to visually effect a "speed line" or to otherwise add interest to the fender area, but the term can be also applied to similar additions over windows or other openings. Few production cars have used fender eyebrows in their styling, probably because of the inherent difficulty in repairing damage from a collision.

1. Punched hood panel/1936 Ford roadster

2. Punched deck lid/1940 Ford coupe

PUNCHING is usually thought of as being synonomous with louvering, but it also refers to the drilling or cutting of holes. In the early days of dry lakes racing, and also in drag racing, cars that ran with fenders often had them liberally drilled with holes to allow the escape of air built up in the wheel wells. The operation, when carried out on structural members such as axles or frame rails, was usually in the interests of weight reduction.

Louvering is the pressing into metal of one or more slots for ventilation, as with an engine compartment where unwanted hot air is allowed to escape. No metal is removed in louvering. The subject panel is placed between male and female dies which are then forced together under great pressure. The pressure tears the metal when one side of the area is raised, and the other is lowered, thus forming the familiar aperature. Used on race cars originally, thus later becoming synonomous with speed, louvers add eye-appeal to custom cars and louvers have been punched into virtually every panel—from those of Model T's to late-model vans—that can be manipulated into a press; fenders, hoods, decklids, even turret tops and quarter panels have felt the bite of the louver press. On panels too large to be worked into a press, or where shape prohibits it, a small piece of metal can be louvered as desired, then the panel riveted or welded into a cutout in the larger panel.

1. First production fadeaways/1942 Buick Roadmaster

2. Custom fadeaways/chopped, channeled '41 Ford

FADEAWAYS applies to the addition of tapering sheetmetal panels to visually extend the line of the front fenders. General Motors had pioneered such a treatment on their 1941 Cadillac Fleetwoods. The fenders extended beyond the cowl to the front door line, then a cap was added to carry the fender back to about the midpoint of the door. For '42, the treatment was passed down to all GM cars, but on some model Buicks, instead of a short door cap, the panels stretched clear along the body.

The fadeaway treatment caught on with customizers who formed similar panels that would look right on many other cars, but was handled most successfully on Fords and Mercurys from '41 through '48. Before the addition of the new panels, though, the rear portions of the front fenders were trimmed off and reshaped so the fender line would flow smoothly back to the cowl area instead of rounding down. To do the job right, this meant reworking the fenders from their highest point over the wheel openings and on back.

Famed customizer Jimmy Summers hit upon the idea, after World War II, of making fadeaways commercially available and had dies made to stamp the necessary pieces to fit any '42 through '48 Chevy 2-door body. Since these cars used door caps already, the front fenders could remain stock.

In a few cases, fadeaways have been used successfully to cover the seams of a sectioning job with outstanding results. With the advent of the slab-sided appearance beginning in 1949, however, cars built later than this cannot have the technique applied. But a late pickup truck might receive such an operation successfully by fairing a scratch-built fender extension into the door panel.

Minor Touches

Little tricks can add up

by Jay Storer PHOTOS BY ERIC RICKMAN

No matter your hobby interest, there are undoubtedly extremes within the field, involving a few top, wealthy people at one end and the bulk of enthusiasts at the other. In stamp collecting for instance, there are $100,000 stamps and there are 13-cent stamps. There are $100 old cars and $250,000 old cars, and boats, aircraft, cast-iron toys, comic books, guns, match folders and the myriad other things that people collect come in many sizes and price tags. While there are inevitably those collectors or enthusiasts who can afford to indulge their passion to the fullest and invest in the biggest, best or most of something, the average participant must content himself with enjoyment of the hobby on a lesser scale. Such is the case with customizing automobiles.

For every fancier that could afford a Barris car or any of the expensive treatments such as chopping, channeling or sectioning, there were thousands of "regular" guys who got into the spirit of the thing with simple bolt-on parts, a few minor metal changes and a nice paint job, or even the "American racing colors" of gray primer. The important thing was to be involved, which meant that no one would be caught dead driving a totally stock automobile. This minor stage of customizing has really never died, making subtle and inexpensive changes are just as popular today as they were back in the early 1950's.

Obviously, there are a lot of changes that can be made to a car without spending a fortune. In fact, such has been the interchangeability history of the auto industry that many changes can be made without doing any metalwork at all. Just by exchanging the grille, headlights, bumpers or taillights from one vehicle to another, a new look can be afforded your car, and in case of the most popular cars or trucks, custom replacements for these items are readily available through shops or mail-order catalogs.

1

2

FRONTAL ASSAULTS

As the most visible part of your vehicle, the front end probably comes in for more than its share of customizing, and if you're limited in time, talent, money or facilities, you may be wise to start here to make the most of your minor customizing. Traditionally, the headlights have been one of the most popular areas for minor customizing. Back in the Fifties, the trick was to french the headlight rings to the fender for a smoother look and paint them body color, even if they had been chromed originally. This is a job seldom performed on late-model cars, but it's a technique that may bear looking into as a subtle change toward cleaner body lines. Obviously, some late-model cars won't lend themselves to this trick because the headlights mount into the body without a separate bezel. Another problem on some cars is that the bezel isn't made of steel, but of alloy "pot metal." These are castings with a fairly low melting point, and are very difficult to weld or braze. One solution to this problem, although not an easy one, is to make a duplicate of the bezel in sheetmetal, and weld that on. Whatever year of car you're working on, if the bezel is a chromed one, you'll have to remove the plating before you can weld or paint it. If you're very careful with a sander, you can grind this layer of plating off without damaging the thin metal of the bezel, but the neatest way is to take the bezels to a chrome-plating shop and have them strip the plating off electro-chemically. After this inexpensive operation, the parts can be readily welded and painted.

Other headlight tricks that are more popular on modern cars include covers and swapping in new rectangular lights. The latter trick has become perhaps the most popular minor custom trend today. Round headlights seemed appropriate back when cars were shaped like pieces of fruit, but ever since Detroit took the longer-lower-wider look to heart, the round lights seemed out of place on the very horizontal and rectangular lines of most front ends. Years ago, customizers used to use European

1. Many of the new 1978 cars have the recently approved single rectangular headlights, like this new Plymouth Horizon. Customizers have waited years for a legal rectangular headlight.

2. One of the most effective bolt-on customizing tricks has always been a pair of side exhausts, which make the vehicle appear lower and longer.

3. Just by modifying the stock light buckets, customizer Dick Dean was able to add new quad lights to this 1977 Courier. The black housings look especially good with that chromed tubular grille. This trick fiberglass hood should be available soon from Fairway Ford, Placentia, CA.

4. Another effective treatment for achieving a lowered or sectioned look is adding something to the rocker panel. This XKE appears sectioned by the addition of this cut-down Corvair aluminum rocker panel cover. Another idea is to paint the rockers panels and other lower body areas flat black.

3

4

rectangular driving lights on show cars, but they were never legal for the street. But since GM introduced rectangular sealed beams on their 1975 cars, customizers have had a field day. The GM lights have been adapted to just about every type of vehicle, with customized trucks and vans among the most commonly converted. On some cars, the conversion isn't a simple matter, because usually the entire GM headlamp bucket has to be accommodated and you have to have that much space, but most of the trucks and vans have plenty of room in the grille and headlight area. Up until recently, the rectangular lights have only been available as a set of four, two high beam and two low beam, so the conversion didn't work well on those vehicles which had originally come with only two headlights. Combination rectangular headlights have finally been approved by the Department of Transportation, and some of the 1978 vehicles have them, including the '78 Bronco, Fairmont and Chrysler's Omni and Horizon subcompacts. In 1975, the most popular units for swapping were the Cadillac and Monza lights, but now that most cars have rectangulars, you have a wider

1

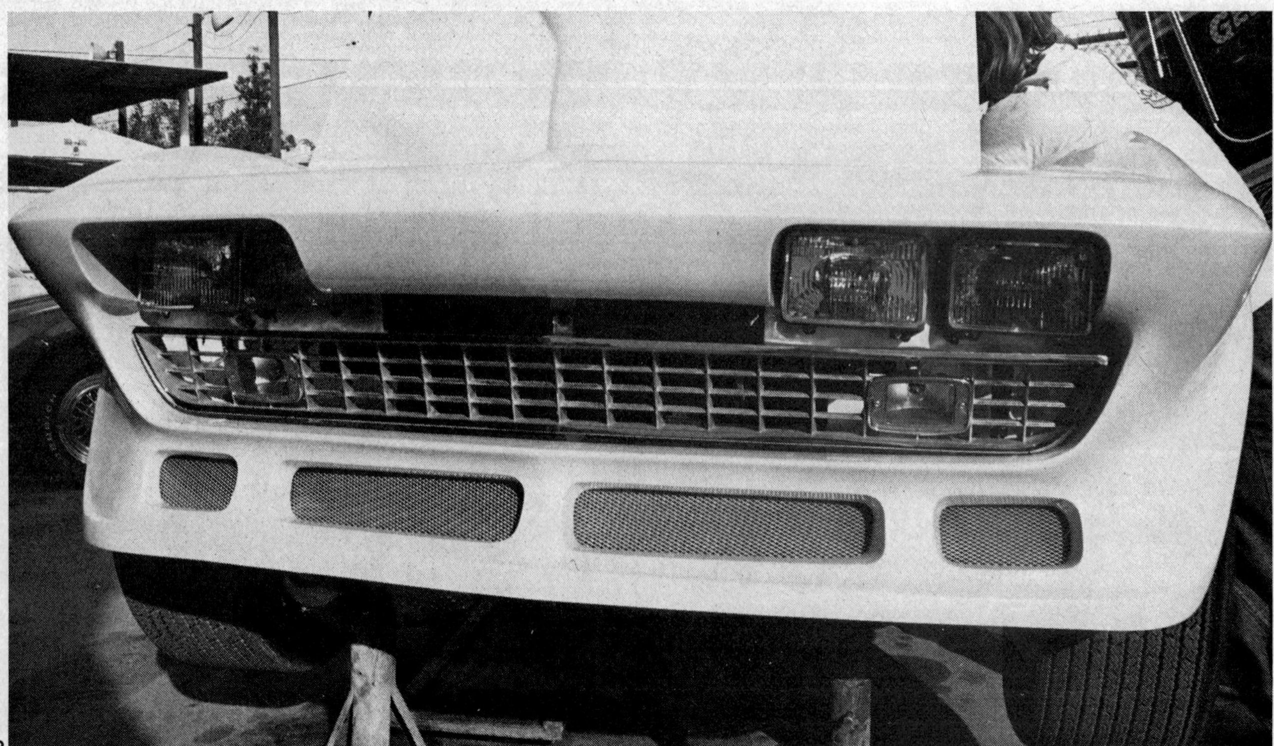

2

1. Spoilers and fog lights or driving lights add a sporty look to any front end. Cars with recessed headlights like this 240 Z can benefit from a set of plexiglass covers in states where covered headlights are legal.

2. The quad rectangular light setups are still quite popular, especially if you want the front end to appear wider. This Corvette by Bruno uses Cadillac lights, eggcrate grille and screened scoops in the spoiler for a total custom front end.

3. The "asymmetrical" look was popular in the Sixties. This can be created simply with a custom grille, or with one of the headlights covered, as at left (the left section of the grille lifts out for night driving).

choice. Whether you go for the quad setup or the newer combination lights (two to a set), there's enough variety in original installations that you can pick an original light and bucket that would fit your vehicle. If yours is a small car, pick the lights and buckets from a new compact, rather than a truck or Cadillac.

Another trick that is both simpler and less expensive (the factory quad rectangular lights cost $90 or more per set) is to fabricate a set of headlight covers, which can create a whole new front end look without touching the factory sheetmetal. It's been a popular custom trick to adapt the hideaway headlights such as used on Mercurys and Cougars to many other vehicles. When shut they look like part of the grille, and you could design a cover for the Cougar lights that would match whatever kind of grille your car had. These electrically opened doors are plentiful in wrecking yards, but like the rectangular lights, you almost have to use them with the factory buckets and hardware they came with, which may be a problem on some applications. Even simpler and with almost the same visual results is to fabricate your own set of removable headlight covers. These can be simple sheetmetal panels that attach to spring-loaded clips around the headlight moldings, and when darkness comes, you just snap them off and stow them in the trunk. The panels can be of metal and painted to match the body, painted flat black and fitted with fins or tubes that make the

3

covers look like extensions of the grille, or you could simply cut some flat covers out of plexiglass sheet. On cars with recessed headlights such as Jaguar XKE's and the Datsun Z car, plexiglass covers are available that "fill out" the headlight opening to create a flush fender line. These really look nice, especially with tinted plexiglass, but in a few states they aren't legal to leave on when using the headlights at night. Check your local vehicle code before you make the installation a permanent one.

The rest of the front end can be updated or restyled almost as easily as changing the headlights. Most common today in conjunction with a switch to rectangular headlights is to change the grille. The custom, chrome-plated-tube grille has been around since the early 1950's and is still popular today. You can make one yourself out of lightweight tubing (such as EMT, or electrical conduit available at big hardware stores) using welding rods for the vertical supports. You can paint the grille or have it chrome plated, and for many trucks and vans ready-made horizontal-tube grilles are available from accessory companies. Your imagination is your only limit in coming up with materials to create your own grille. You can use expanded-mesh metal, plastic egg-crate grilles from overhead flourescent light fixtures, or swap the grille from another car that would look good on yours. The one thing to keep in mind is that whatever type of grille you build, it has to flow enough air through it to keep the engine cool. Using pegboard, for instance, wouldn't be advisable.

Other than the grille and head-

1

1. In terms of bolt-ons, nothing could be simpler than a pair of aftermarket mirrors, like these chromed Talbots. Streamlined plastic ones are popular too, either flat black or body color.

2. This street cruiser features an old custom trick, a sunken antenna, or pair. This is usually done when a power antenna is added. Note also the Metalflake roof and the diffraction tape used within the side trim.

3. Here's a subtle touch. The owner of this Corvette has molded a fairing around his outside mirror, although the glass is still adjustable.

4. Flared fenders to cover wider tires have retained their popularity for years. Rear flares on this Ford Torino were done in sheetmetal with a conduit edge (see our how-to on these type of flares on page 78) to give a factory look, but flares are also available for many cars and trucks in bolt-on fiberglass form.

5. One of the street customizing trends of the last five years has been the "road racer" look, such as on this Mustang, where spoiler, apron, scoops and plexiglass quarter windows are utilized for a Shelby look.

2

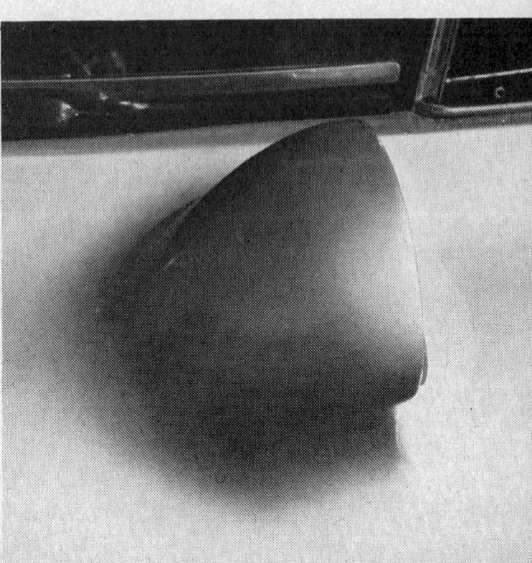

3

lights, the most commonly customized areas of the front end would be the hood, bumper and lower gravel pan. In keeping with the road-racer look that most street enthusiasts are after today, spoilers or air dams are often added to the lower gravel pan to make the car appear lower, and the addition of driving or fog lights under the bumper completes the "rallye" look. Not much else can be done with the gravel pan, but depending on your car, cutting in some air scoops or molding in a set of accessory lights might look good.

You wouldn't think that much could be done to the bumper of a car, but there are some subtle modifications. The easiest trick is to remove the factory bumper guards, if your car is so-equipped. On many cars, the upright bumper guards break up the low-'n-wide look customizers try to achieve. You can either fill in the resulting bolt holes in the gravel shield with body-color-painted carriage bolts with flatheads, or weld up the holes entirely. On the bumper itself, you can fill the holes with plated bumper bolts for a factory look. In some cases, removing the bumper guards exposes the fact that your bumper is not one-piece and the guard covered the seam. This requires some expense, but these bumpers can be cleaned of chrome (near the seams), welded together, ground down and replated. Voila, a smooth, one-piece bumper. To carry this clean-up process one step further, you can "french" the bumper bolts. There's nothing obscene about this popular trick from the '50's, you just weld studs on the back side of the bumper to replace the bumper bolts. The stock holes on the face of the bumper are brazed or welded up and the bumper replated. This makes the bumper completely smooth, with no bolt heads showing to the outside.

If you're brave enough to delve into a little metalwork, the hood offers several approaches for adding a new look. Removing the stock insignia and trim and filling the holes (by welding, not Bondo) is a routine operation and one that is just

4

5

as popular and applicable for today's cars as it was in the customizing heyday. A DA'd custom car owner of the Fifties was once asked by a nonenthusiast why he spent so much time and money to have every bit of trim removed that Detroit had probably spent hundreds of thousands of dollars to develop. This early-day Fonzie's unhesitating but always-cool reply was simply, "Heeeey, man, that stuff (chrome) used to tear up my chamois!"

Another popular hood treatment of the Fifties that has practical applications today is louvering. Twenty years ago there was a louver press in every large town, so popular had this fad become. When it started, some unknown hot rodder cut out a section of louvered sheetmetal from an old gym locker and hammer-welded the panel onto his car. This finally caught on to the point where no custom shop was complete without a louver press, and every part of a car was subject to the treatment, particularly on hot rods. There was a practical reason for louvers, at least when applied to the hood of a car, because they allowed hot underhood air to escape and kept hot-running flathead Fords cooler while cruisin'. With the high engine compartment temperatures our smogged engines live with today, the louver is back, and the current nostalgia trend has also helped to keep the presses going. Louvers can be various sizes from approximately 1½ in. wide to about 4 ins. wide, and laid out in a variety of patterns and rows.

Carrying the engine cooling a little further, hood scoops have always been well received, especially among the street performance crowd. Getting cooler air into the engine compartment means better performance, and a properly located scoop can provide cool air directly to the air cleaner. Unless you're handy with metalwork, you'll probably have to settle for using a fiberglass hood scoop, of which a wide variety are available. Some can be bolted on, others have to be fiberglassed on and bodyworked before painting. Some cars, (the sporty compacts such as Vega, Monza, Capri and Toyota for instance) already have a hood bulge in the stock metal that needs only a simple cutout in front to make functional.

BRINGING UP THE REAR

Other than paint treatments, which are covered in a separate chapter, there aren't many minor custom tricks that can be applied to the sides of your vehicle. Bolt-on equipment, such as custom tires and wheels, side exhaust pipes, trick outside mirrors, mirror fairings, special chrome trim, and lowering of the suspension is about the limit of minor customizing that affect your profile. One of the most popular custom treatments today is flar-

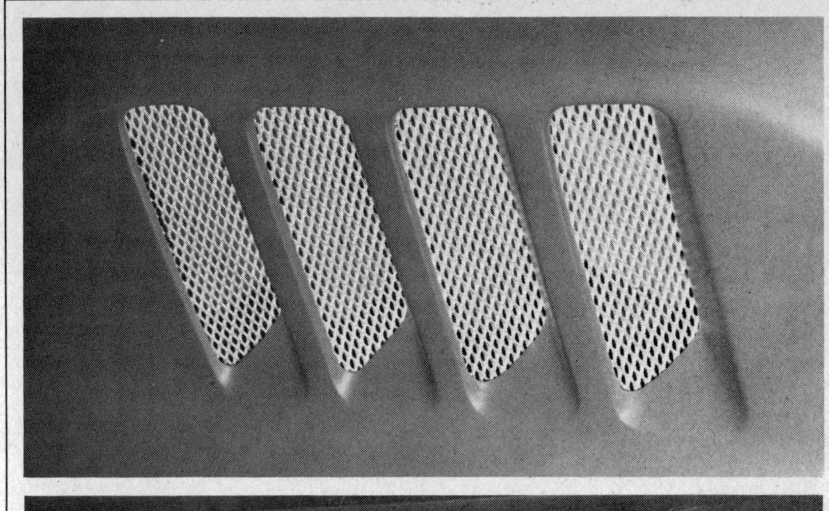

1

2

ing of the fenders, for the road-racer look and to stay legal by keeping your oversize tires covered. There are dozens of custom wheelwell flares available in fiberglass that you simply pop-rivet in place (see the fiberglass chapter), but otherwise the job is one that requires metalworking that keeps this out of the "minor" category.

At the rear of a mild custom, we find some of the same techniques applied that are used on the front end, especially with regards to the bumpers. Decklids seldom come in for much customizing (unless you have a VW or Corvair and want to add a scoop or cooling louvers to the decklid), other than routine emblem removal. We have seen a few examples though, of cars with thin taillights molded into the decklids to achieve a look like the newer Eldorados and Toronados, which have such lights as a safety feature. The most common trick for decklids is probably the bolt-on spoiler. Both Ford and Chevrolet have these in stock for their Boss Mustangs and Z-28 Camaros, and these can be adapted to many other cars. Also, there is a large variety of aftermarket fiberglass units available, either for bolt-on or bond-on attachment.

Taillight swaps and exhaust outlet treatments are two of the easiest and most effective changes you can make at the rear of a vehicle. Use your imagination here. When you're driving in traffic, study some of the taillight designs you see in front of you (watch the road too) and select a pair that would look right on your car. Cars with a fairly flat panel between the decklid and rear bumper have the most possibilities for swapping taillights, but even if your stock taillights are mounted in cast pot-metal pods at the rear of your fenders, there still may be a cleaner unit from some other car that will fit. Thousands of designs pass by you every day on the road, just keep checking them out.

There are a number of ways you can go with exhaust treatments, depending on the look you're after. If you want the performance look, you'll definitely be switching to dual exhausts, and any of the many chromed accessory exhaust tips can complete the picture. Or you could route the exhaust to come through the rear gravel shield (ala the GT Mustangs) and build custom chromed rings to mount to the panel around the tailpipe tips for a factory appearance. For the sporty look, you can go for the treatments you've seen on many expensive sports cars. Either run the single or dual tailpipes right up the center of the car, or cut out a hole on one side of the gravel shield at the bottom and group the pipes there for the asymetrical look. Multiple-pipe exhaust tips can be purchased at sports car accessory stores, or you can make up your own from exhaust tubing, weld them together in the desired pattern, and have the piece chrome-plated or paint it with black wrinkle-finish paint. Take a look at some of the foreign cars and see how the tailpipe is usually set into the bottom edge of the gravel shield. This looks professional when applied to a custom, especially if you finish off the opening with a beading or rolled edge. A local sheetmetal shop can probably roll a nice edge for you if you bring them the panel off the car, or you can weld or braze some welding rod around the opening to form a bead.

You'll find many other customizing tricks throughout this book, some advanced and some easy enough for a beginner, particularly in the paint section, where significant appearance changes can be made more easily than with metalwork. What you do in the realm of minor customizing is limited only by your imagination, talent and facilities. We've illustrated here some food for thoughts on what anyone can do to effect subtle changes before advancing to even more creative work.

3

4

1. Expanded mesh metal is available at most hardware stores. It makes an excellent screen material for custom vents and scoops. On this Corvette, the mesh was fiberglassed in behind the front fender air vents.

2. Taillight swaps have always been one of the easiest custom tricks. A pair of Camaro lights were added to this customized Corvette, then side markers were used in the lens centers for further disguise. Through-the-pan exhaust tips (stock on this Corvette) lend a performance look to some cars.

3. That trick reverse scoop on the front fender of this van wasn't built from scratch. It's a vent from a Pontiac Trans-Am Firebird molded in. Accessories and body parts from many different foreign and domestic cars could be adaptable to your vehicle.

4. Another Corvette, this 1963 model features shortened bumpers at rear, flanked by extra stock lights on each side and extra exhaust tips.

The Plastic Revolution

The coming of "cold" fillers

by Jay Storer PHOTOS BY ERIC RICKMAN

One of several memorable scenes from the Mike Nichols film The Graduate was when Dustin Hoffman was taken aside by the husband of a neighbor and given some well-meaning advice. The man's one-word summation of where the money was to be made and where the world was headed was "Plastics!" Though the scene was a humorous one, that advice over 10 years ago was basically sound. We can see the depth to which plastics technology has penetrated our everyday lives. We find these man-made materials in everything around us, our cars, furniture, homes, clothes, appliances, highways, etc. Plastic substitutes have been made available for just about everything but sex and food.

While the incredible usefulness of plastics materials today would have made good investment advice some years ago, there was a time when plastics as applied to automotive bodywork had a rather low status, at least with customizers. Fiberglass, or "glass-reinforced plastic" as it was called, had been used during World War II as a substitute for some metals in boat construction, and consisted of glass fibers or cloth made rigid by the addition of an epoxy resin. Total acceptance by the public was still a decade away though. In the early 1950's, there appeared on the market fiberglass kits specifically aimed at the auto hobbyist. Some were used in the Eastern states to repair rusted out rocker panels etc., and a few were tried for customizing jobs, but most bodymen and customizers wouldn't have anything to do with these "cold" repair kits. They preferred the tried-and-true methods. You could as soon step on the man's "blue suede shoes" as propose to take away his torch.

The Korean Conflict had the same effect as other wars on the price of strategic metals, and the cost of lead went up as the government took most of the supply. So those industries that used a lot

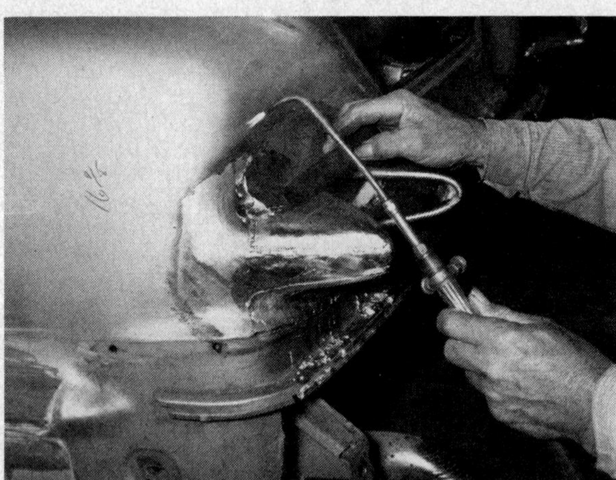

1

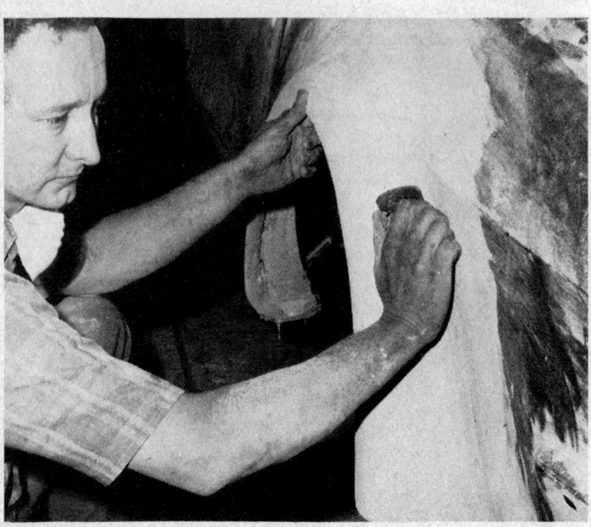

2

3

of lead, such as plumbing and body repair, took a harder look at substitute materials. So many small companies started out in the '50's making compounds that it's hard to say who exactly gets the nod as the absolute first. One of the first at least was Joseph O'Donnell. He had been marketing a repair kit called "L-Bondo," which was well received in the plumbing and boat repair fields. He hit on the idea of mixing the kit's resin compound with some talc filler and using it independent of the glass fibers. This he called simply Bondo, and it became the forerunner of the many plastic body fillers we use today. It worked so well as an auto body filler that in 1955 he started the Bondo Corporation in Northford, Connecticut, in order to market it. Other companies soon joined the fray, such as Swiss Laboratories of Akron, Ohio, with their Sno-White brand. Swiss had been in the body lead business, and the plastic filler was a natural product of their research for a cheaper material. The Sno-White brand is still around today, evidence of the growing acceptance bodymen found for such easy-to-work materials. Another early material that found some acceptance was called "resurfacing emulsion," which users found to be more flexible than some of the other early plastics.

Joe Bailon of Hayward, California, was one of the first of the big-name customizers to use plastic fillers. The Taylor and Art Plastics Company of Oakland, California, had been marketing a fiberglass body repair kit called the TAP-Kit, which they aimed at the do-it-yourself customizing market through ads in enthusiast magazines. In a 1957 issue of Rod & Custom, Bailon was featured adding some plastic fins to an early Studebaker using a TAP-Kit, and he has used the man-made materials almost ever since, except when a customer specifically called for the use of traditional lead. Joe found after using a few of these repair kits that he was allergic to the itchy fiberglass dust. He quit using the kits, but found the resin-based filler that came with them made bodywork easier. Why not, he reasoned, just do the metalwork as always with sheetmetal and welding rod? Only now, he finished the job off with plastic filler and saved all of that tedious lead work.

Many of the other famous customizers of the 1950's didn't start using plastic fillers until the early Sixties. The company started by Joseph O'Donnell made such a hit with their product through regular bodyshops and for do-it-yourselfers that Bondo has become the generic term for plastic fillers even to this day. The Bondo Corporation merged with the Dynatron Corporation in 1973 and continues to make the filler.

Plastic fillers had some problems when first released to the market. Some were too hard and brittle, some wouldn't bond properly to the metal, and some cracked out in cold weather. These factors plus a conservative reluctance to change kept the early customizers away from using fillers in their expensive and show-oriented jobs, but as the products improved through the 1950's, the plastics became more common. By the early 1960's, lead was still being used only by a few of the remaining "old masters," and then often by request only.

The use of the fiberglass products and plastic filler led to a greater proliferation of customizing because they were so easy to use. Little or no torchwork was necessary with the majority of minor customizing jobs, and this really brought customizing home to the garage/backyard enthusiast. Even a more advanced job could be set up by the enthusiast, brought to a body shop for just the welding, and then brought home again to be finished in do-it-yourself fashion with the easy-working plastic fillers, or "cool-lead" as some called it. The plastic products not only lowered the cost and time required for customizing by pros, it also opened creative doors for the host of amateurs around the country, contributing to the full flowering of the customizing field in the mid-Sixties.

Today, plastic fillers are the mainstay of both production bodywork and customizing, in addition to the myriad of custom parts available in fiberglass ready-made. The use of plastics today—when we have everything from custom replacement fenders to complete custom bodies as near to any of us as a mailbox—is a major part of the current rebirth of creative customizing.

4

1. From the early days of the coachbuilders through the customizers and right up to the Sixties, lead was the exclusive filler medium. It wasn't cheap, took a great deal of skill to apply properly, and was tough to use on vertical surfaces, but it did the job for those who mastered it.

2. Eventually, plastic fillers improved to the point where they became the rule rather than the exception, both in customizing and regular bodywork. They could be applied on any surface, worked faster and easier, required no torch, and were in fact the only way to repair fiberglass vehicles like the Corvette, Avanti and others.

3. The early to mid Fifties saw the emergence of lead substitutes such as fiberglass and epoxy-type fillers. The TAP kits were among the first plastics aimed specifically at the customizer and automotive hobbyist.

4. Here's a graphic comparison of the difference in application and finish techniques for lead and plastic filler. At left are the torch, beeswax, paddles, lead sticks, tinning compound and steel wool needed for leading, compared to the hardener and plastic spreader for applying flexible plastic fillers.

HOW-TO: HEADLIGHT CONCEPT

PHOTOS BY ERIC RICKMAN

Creativity in customizing can take many shapes and forms, either through the mind's eye of the customizer, or through the sometimes bizarre requests of a shop customer. Either way, it is up to the bodyman to carry the treatment through to a satisfying finish whether he's creating a whole new shape or merely reworking a detail line or body crease.

Our scurrying photographer happened to drop by Customs by Eddie Paul just as Eddie had begun a major facelifting of a pseudo '35 Ford pickup, following suggestions laid down by the truck's owner—who hastened to add that early-Ford restoration buffs shouldn't be upset by his seeming violation of the resto buff's creed to keep everything stock. The subject, you see, was an assemblage of various year Ford parts and other model components rather than an unspoiled pickup. The project involved the forming of a new grille, although it was designed to somewhat resemble the original counterpart, altering the grille shell to accept the new barred inset, and to employ quad headlights. As noted above, customer's requests can sometimes be termed bizarre, but in the end the frontal effect is much more pleasing than a verbal description suggests. Some of our staff members have chanced to see the quaint little hauler bustling along Los Angeles area freeways, and all agree that the front end is tastefully unique.

The carrying-out of the project was rather straightforward, and the quad light assembly came from a '77 Monza.

If you have ever purchased a set up over the counter at an agency, you know that each individual part of the quad lights have to be ordered by separate part number. Eddie Paul, however, can supply the entire assembly in kit form.

1 This is what faced the customizer; an odd mixture of early Ford parts all hung on a Chevrolet-powered '35 pickup. Project had quad lights, grille.

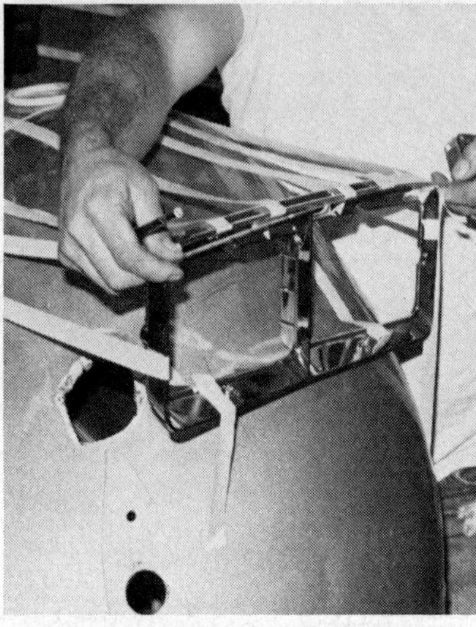

2 Monza quad light bezel was taped in location determined by "eyeballing," so bodyman can first stand back and view the concept.

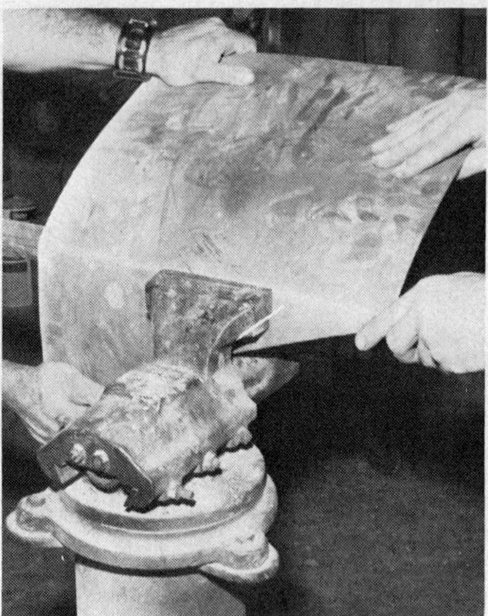

3 Housing of sheetmetal was cut out from pattern made over taped-up bezel, then bent as required in a vise. The metal is same gauge as fender.

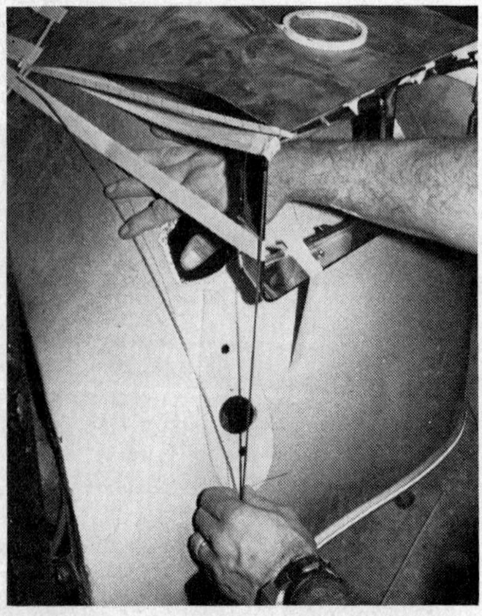

4 With part of light-covering shroud taped in place, welding rod is formed to outline of yet needed inner side. Sheetmetal was shaped next.

5 Monza light bezel perimeter is traced on piece of sheetmetal, this shape is then cut out and the panel thus becomes form of shroud opening.

6 Rod is heated and bent to conform to panel's outline, is temporarily tack welded to it to hold the shape. Rod will form shroud opening edge.

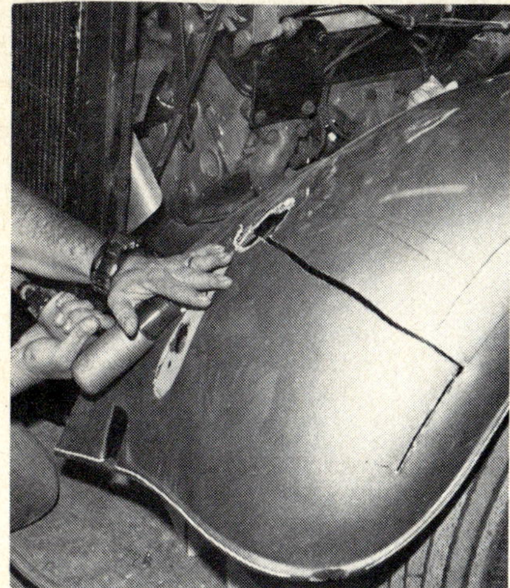

7 Part of the original fender has to be removed to let headlight bucket fit. Air chisel is the best bet with the heavy, original fender material.

8 The shroud is clamped in position to check location and fit, and to make certain that the lamp assembly has enough room for mounting.

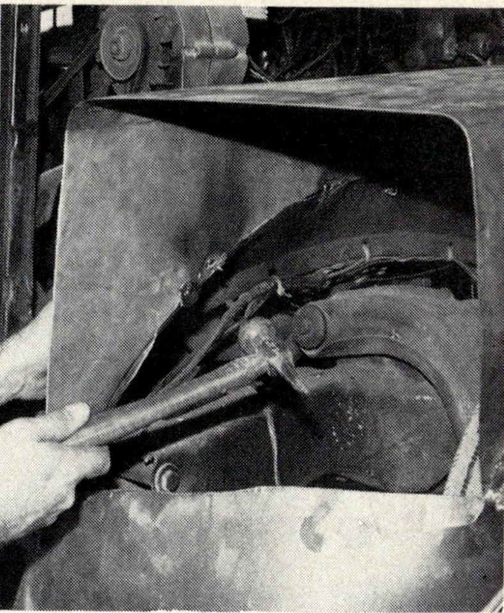

9 Tack welds are made at intervals; edge of fender cutout is hammered up inside lower shroud edge to form a double thickness for strength.

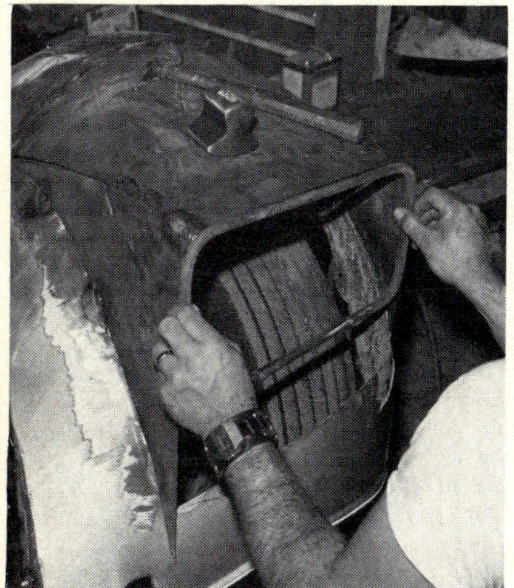

10 Rod formed around temporary panel is checked for fit. Same panel is used again to form the rod for the opposite fender.

11 Now it's beginning to look like something, as Monza lamp bucket is checked in its recess. Welded-on screw tabs to be added.

12 Welding is completed, and all the seams and surrounding metal have been ground down to clean metal. Now, do second fender.

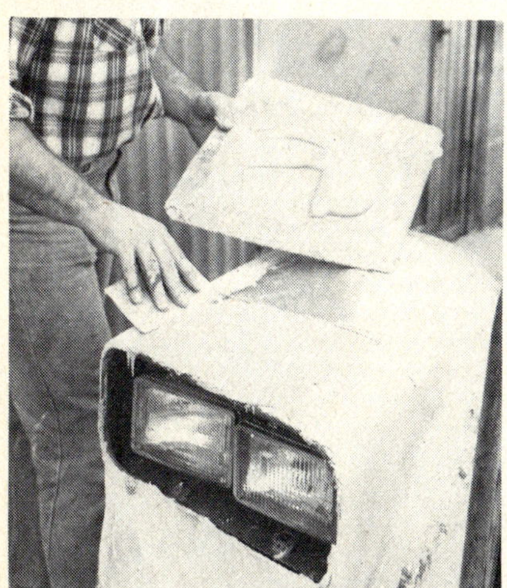

13 Bondo is mixed on a pallette, applied sparingly to fill low spots and, cover the welded seams. Filing will show where a second coat is needed.

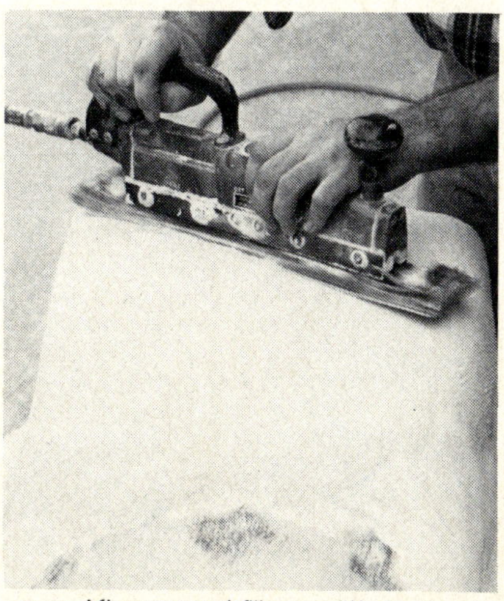

14 After second filing, a jitterbug sander is used to smooth everything out and get rid of file marks. From here on out, it's straight finishing.

15 Eddie also built a custom grille from light tubing. Other parts of the truck need attention yet, so owner drives it in primer temporarily.

Fiberglass Put-Ons

Though pioneered in the early Fiftes, glass-reinforced plastic is still climbing in popularity.

by Jay Storer

Fiberglass products as a substitute for metal in automobile bodies have been with us for a lot longer than most of us realize. During World War II in fact, some fiberglass hulls had been built for minesweepers due to the shortage of metals. At that time, the material that resulted from mixing resin, hardener and shredded glass fibers was called "glass-reinforced plastic." One of the most important companies in the early development of the glass fibers used in GRP was the Owens-Corning Co., who later trademarked their product as "Fiberglas," which eventually became the generic term for the material. The U.S. Rubber Co. was also instrumental in the application of boat-building technology to custom automobiles. Their plant in Naugatuck, Connecticut produced the "Vibrin" polyester plastic used by many of the early boat builders, and when they learned that one of their customers had built a car body from their materials, they got behind this potential new sales outlet in a hurry. Glasspar Boats of California had built a one-off body for a Jeep chassis in 1951 which spread the exciting news of durable, rust-free auto bodies nationwide when it appeared in Life Magazine in 1952.

1. First of the name customizers to build one-off cars out of fiberglass was Ed "Big Daddy" Roth, shown here building up plaster over plywood to form the "buck" for the Mysterion car in 1963. When he was satisfied with the shape of a project, he would sand it smooth, paint it and make a mold, from which he could make other bodies.

2. Another famous show car of the Sixties that used fiberglass to good advantage was Steve Scott's creation called the Uncertain-T. Shown here in raw form, the simple-yet-wacky lines showed the creativity of 'glass.

3. Many customizers were really first introduced to fiberglass when they had to do a Corvette the first time, and learned the creative possibilities opened by this material they had formerly looked down upon.

1

2

3

4

4. The Sixties was a "dune buggy" generation. Thousands of fiberglass bodies, all different yet alike, were sold for shortened Volkswagen chassis. Though the term wasn't bandied about at the time, this was really "recycling" of old VW's with rusty or dented bodies.

5. Today, your VW fancy can be turned to either the practical or the purely sexy. A case of the former approach is this VW flatbed kit from Domus.

6. Fiberfab is one of the oldest names in the kit-car field, and their latest entry, the Aztec 7, shows why their success has continued. This was a VW?

7. The use of fiberglass for bolt-on customizing was first explored when companies built flared fenders for those VW's that hadn't been turned into complete dune buggies. Now a lot of today's customs use bolt-on 'glass.

8. Some kits have become so popular that owners have formed their own clubs, like this gathering of Sterling cars. Use of engine adapters allows a choice of non-VW powerplants, too.

9. Two things have been said over and over about fiberglass, it won't rust and you can duplicate just about anything with it. The scarcity of restorable old cars has led to duplication of many desirable bodies in 'glass, such as this '32 Ford coupe by TSR.

This was the point at which Chevrolet got interested in this new material, thinking it the perfect medium for producing a limited-edition sports car they were planning It was introduced shortly afterwards as the Corvette, beginning an important era in automobile history as well as a milestone for GM, who recently celebrated the 25th anniversary of that sports car.

Other people had experimented with plastic bodies for cars in the early Fifties, but none had GM's wherewithal to make serious production a reality. Most of these cars were one-offs or of very limited production, but since they were almost all unusual sports car bodies, they heightened public enthusiasm for the materials as well as the designs. It became readily apparent to enthusiasts in the Fifties that this was the easiest route to building a custom car (not customized, but a wholly new body). Who knows how many backyard stylists created sporty bodies for their Crosley, Willys, or cut-down Ford or Chevrolet frames as fiberglass technology spread. Customizers were still using steel, welding rods and lead for their restyling projects, but by the mid-Fifties fiberglass repair kits were already being marketed and aimed at the customizing field.

Customizers were slow to accept these new materials as replacements for the traditional methods, but once they were convinced to try them, they usually welcomed the creativity that fiberglass and Bondo made possible. By the early Sixties, some of the top names were starting to use fiberglass as a medium when building far-out cars for the show circuit. One of the first to make a name for himself in this kind of work was Ed "Big Daddy" Roth of Maywood, California. Roth saw the creative possibilities of fiberglass long before anyone else

5

7

6

8

9

and became famous for his free-form-bodied "wierdo" cars such as the Beatnik Bandit, Mysterion, Road Agent, Outlaw, and Rotar. To him, customizing with fiberglass instead of steel was as different as making animated cartoons is from making films. However you create your original "buck" before laying fiberglass, your horizons are limitless because changes can be made as you go along. When working in metal, changes-as-you-go are time-consuming and expensive, and when everything is welded solid subtle changes are the hardest to effect. Working in plywood and plaster, Roth would build his cars like a beatnik sculptor, and when the carving was done, he'd reproduce the shape in fiberglass, fit it to a special show chassis and top it off with some very freaky (for the period) paint.

The role that fiberglass materials have played in the customizing field has increased drastically from those early days when most customizers refused to work with the then-new materials. Today the majority of customizing involves fiberglass in some way. Those top customizers who still do one-of-a-kind work for customers who can afford to pay the price such talents demand today usually work in steel rod and metal. Corvettes and other fiberglass cars are obvious exceptions. But there aren't as many customers today for this type of work. Most really radical customizing jobs cost in the thousands of dollars, and to make such a job pay off, most customizers will finish a neat machine and take fiberglass molds from it. As an example, let's say a customer comes into a shop with a popular make and body style of car, and he wants the latest look in racing-style flares and spoilers. The customizer gives him a reasonable rate on the job because he knows he can recoup later from the parts. He customizes the car in steel and finishes it off with glossy paint. Then he carefully lays up enough fiberglass material over the modified areas of the car to make a sturdy mold, and does this for both side of the car. Then he reinforces these female mold parts and can then lay up new fiberglass fenders, hoods or whatever in these molds. He can offer these bolt-on pieces to other owners of the same type car for a frac-

1

2

1. Spoilers and wheelwell flares are the most popular fiberglass body parts for customizing, and the sporty cars like the Datsun 240Z receive the lion's share. These bolt-on front fenders and bond-on rear flares are from Carl Green Enterprises. A side benefit of the custom parts is that they can salvage a rusted-out car.

2. Dick Barbour Datsun is one of the many companies making fiberglass for the popular 240Z, including flares and front and rear spoilers.

3. Other elements available from Barbour include these rear flares, skylight roof panel, rear window shade and Porsche-style whale-tail.

4. This trick extended front end from Barbour features an apron in front of the hood, no bumper, molded in flares, and plexiglass headlight covers (illegal in some states).

3

4

tion of what it would cost to have the work repeated by the customizer on another car.

FLARES AND SPOILERS

So customizing today, since such a large part of what people are doing to street machines is in the area of flares and spoilers, is almost a bolt-on proposition. There are literally hundreds of small and large companies making production fiberglass pieces for customizing everything from vans and trucks to mini-cars.

The market grows daily as new companies pop up making parts for a model of car that no one else had covered up to that point. If you own a 1975 Widget, relax, you may not have to customize your car at all. Eventually, there'll be either a fiberglass kit for restyling it, or even a complete replacement fiberglass body.

The "kit cars" are perhaps the most drastic examples of customizing with fiberglass. The phenomenon began in the early Sixties when Bruce Meyers designed and built his first fiberglass "dune buggy" body. Intended to fit on a shortened Volkswagen floorpan, these bodies allowed just about anyone to build rather simply a clean-looking, roadster-type vehicle that was lighter and more nimble than a stock VW and had the necessary clearance for big off-road tires and wheels. They became something of a national fad in a very few years, and scores of fiber-

5. These Barbour conversions for new rectangular lights are a bolt-on, replacing the stock front fender caps.

6. If you're really into the racing look for the street, Barbour also has this "dam" type spoiler with the small wing, called an "aerodynamic snout."

7. Any speed shop owner will tell you some of the fastest moving items are fiberglass flares and spoilers for vans. Like most, these Karvan parts can be riveted on or molded on.

8. Trucks are the latest outlets for customizing with fiberglass. Hickey Enterprises makes this drop-center 'glass hood and front spoiler.

9. Another Z kit, This set of rear flares, replacement fiberglass front fenders set up for rectangular lights, and the whale-tail are from Jim Cook.

glass companies sprang up making copies of other people's buggy bodies. Unfortunately, ripoffs have always been a part of the fiberglass business. The materials are so useful for duplicating shapes that many companies got their start in business simply by buying an existing kit body and making a mold from it, perhaps making one or two minor changes here and there before reproducing it. The dune buggy craze finally was overkilled because so many companies were making them that the price was low and the market flooded.

But the Sixties also saw a new direction for kit-car bodies, the sports and replica markets. Companies

1

2

1. Mini-trucks come in for their share, too. In addition to the usual flares and spoilers, there are also step-side bed conversions like this one by California Step Side. The bed is steel, with fiberglass fenders and cab end caps (arrow), steel steps.

2. The Hardy LUV kit features full IMSA racing styling with low air dam up front and four widened fenders.

3. When race-car chassis builder Don Hardy built a swap kit for putting Chevy V-8's into LUV trucks, he made it a complete package by offering the body parts to make Chevy's mini as fast looking as his kit performed.

3

like Fiberfab, Astro and Kellison started making swoopy sports car bodies that also fit on Volkswagen chassis, turning the German sow's ear into something of an Italian silk purse with a race-car touch. Some of the early kits, it must be admitted, weren't the tops in fiberglass quality, and few could be "assembled easily at home with basic hand tools in a few weekends" as the advertisements claimed. But they offered a new look (the dune buggies were almost as common on the road in some regions as stock VW's) and many of the sports car kits were designed for a stock Volkswagen floorpan. Not having to shorten and reweld the floorpan such as a dune buggy kit required

4. Hooker is a familiar name when you think of exhaust headers, but they're also into fiberglass now. This is their package for the late Camaro, reminiscent of the IROC cars.

5. Up until just last year, Ford didn't offer the Pinto "Cruisin Van," but owners of earlier Pinto wagons can make the conversion from family wagon to trick delivery with kit from Wike Fiberglass Co.

6. Maier Racing has a variety of fender flares, scoops and other body parts for Mustangs. This '66 features racing apron, hood, flares, side scoops, and spoilered 'glass decklid.

7. Owners of early Mustangs can get that Shelby GT-350 look with parts available from Maier Racing. Shown here is their fiberglass hood with scoop and racing-style front apron with an integral front spoiler.

8. Fiberglass treatment for the Pontiac Firebird includes flares and spoiler ala the T/A, and the Hooker hood allows a hi-rise intake system.

4

5

7

8

6

meant a lot more people could tackle the conversion at home. That may also have been a negative factor in the development of fiberglass as a customizing medium because many people who were convinced to buy the neat-looking kits didn't have the time, talents, money or whatever to finish one of these cars, thus giving the field of kit cars a bad name.

Our marketplace has always been controlled ultimately by the consumers though, and eventually consumer demands for quality forced the production of better kits. The realm of VW-based kit cars is enormous today. You can buy everything from freshly designed sports cars that look like factory styling exercises to little pickup trucks to replicas of classic and antique cars, all designed for that same 95-in. wheelbase Volks chassis. There are also a few kits offered today that come with their own tubular steel chassis designed to accept a midships V-8 engine for genuine GT performance in a homebuilt street machine.

1

2

1. Hooker's biggest project to date has been their Vega package. The IMSA look in full bloom tops off their engine swap kit for V-8 Vegas, but you can put the same body parts on your own Vega, even with a four!

2. Everybody's into the IMSA look today, even the lowly Datsun B-210 econocar looks like racing material with this kit from Mansfield Plastics.

3. There are literally hundreds of different scoop designs available in plastic or fiberglass for bolting or bonding to your hood. These are some of the styles made by A&A Fiberglass.

4. A new vehicle doesn't have to be on the market long before someone is pulling molds from it. Automotive Development already has a complete IMSA-look kit for Ford's new Fiesta.

5. Here's how the prototype AD parts looked before painting. You'll need some wide tires to fill those fenders!

6. If you've tried mating fiberglass parts to metal bodies without much luck, this is for you. Available from Carl Green Enterprises, this epoxy adhesive bonds fiberglass to metal positively and permanently. Perfect for adding a scoop or bonding on a pair of fender flares or spoilers.

7. Something new in recent years has been the development of flexible plastic flares and spoilers made of Urethane like these van parts from Automotive Accessories. They're more chip and crack resistant.

8. You say the factory Chevrolet "Mirage" package for the Monza isn't racy enough for you? How about the IMSA/street kit from Dekon Engineering. It includes 2 spoilers, nose panel, four fenders, door entensions and a racing styled fuel filler cap.

9. Carl Green Enterprises offers a variety of parts for AMC Pacers, such as flares and spoilers, and window blankout kits for wagons and sedans.

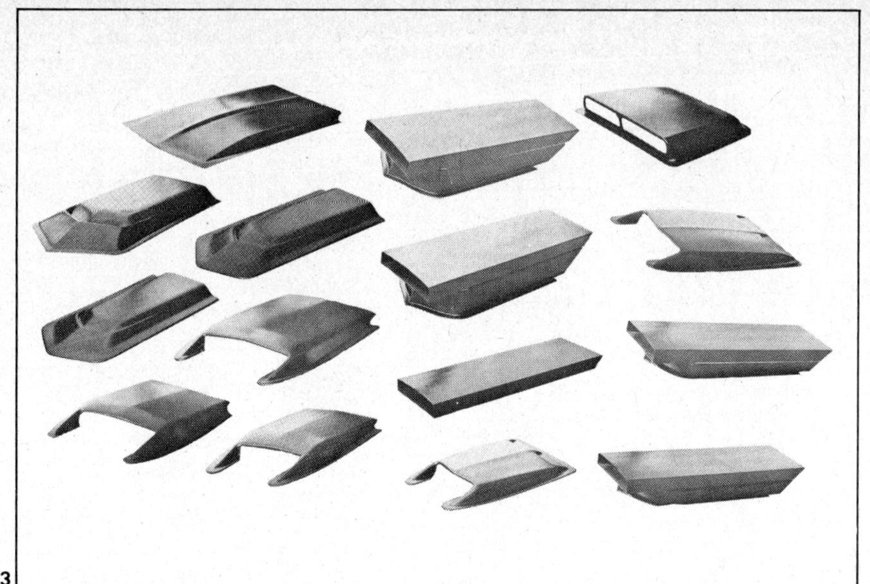

3

Obviously, there's no limit to what can be done with GRP or fiberglass. As Detroit is forced to lighten their cars further to comply with Federal economy standards, we'll see more and more plastic and fiberglass in our cars. Serious experimenting has already been done with a new type of plastic called "GrFRP," for what is actually Graphite-Fiber-Reinforced Plastic. Even stronger than traditional fiberglass, the new material is being looked at as a substitute for steel not just in body and interior parts but also in chassis and suspension members, even springs. In the future, not only will you be able to customize-by-mail by ordering trick fiberglass body parts, but maybe your next trick intake manifold or "mag" wheel may be plastic, too!

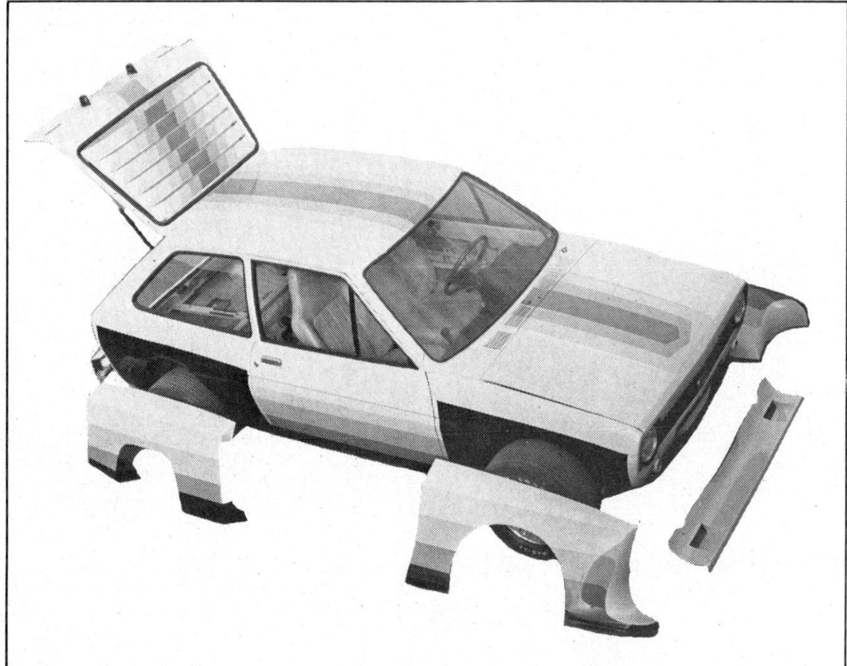

4

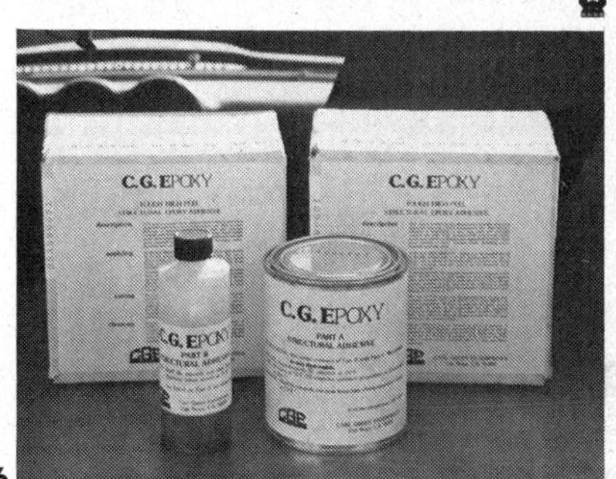

6

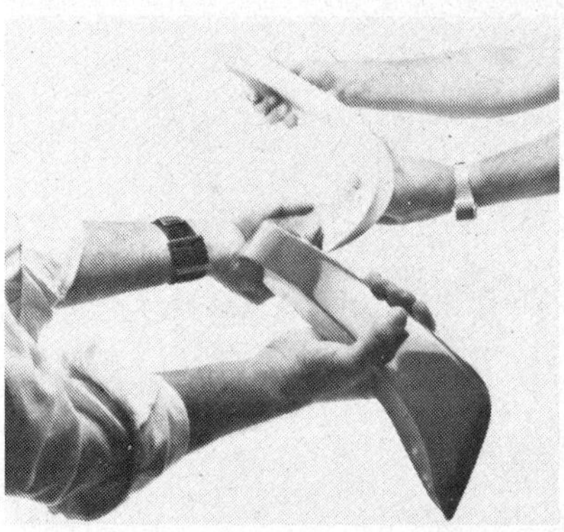

7

5

8

9

The Corvette World

Just scratching the surface of custom-made add-ons for America's only sports car.

If there is a single make of car that receives, on a percentage basis, more personalization than any other, it has to be Chevrolet's Corvette. Rare is the Vette of any of its 25 years that hasn't had at least the addition of custom wheels. Watch the roads and try and spot one that's box-stock. The reason for this is simple; the Vette has never enjoyed production model runs of more than 46,500 units (for 1976). And while that would make a long string of traffic if they were placed end to end (but don't try and pass them), it's peanuts by Detroit standards, so consider the marque rare. Now, since the Vette is purely and simply a sports car, aimed at the more-or-less affluent younger set, few such buyers would ever think of leaving their car as-is.

Such a target is the Vette for all manner of add-on items, it's little wonder that aftermarket firms near and far have come up with spoilers, air dams, flares, scooped hoods, front and rearend "clips", and so forth, to alter, change, improve (in some case, even downgrade) and otherwise help you create a one-of-a-kind sportster. In fact, it's a safe bet you could remove the entire factory body from an any-year Vette and build a new one by bonding together the various pieces offered in any decent Vette aftermarket parts catalog.

Naturally, most of the larger custom panels and other add-ons available for Vettes are made of

1 Each of the leading Vette parts suppliers and many of the customizers who dabble with glass add-on products have their own favorite brand of "goop"—or they formulate their own secretly. Bruno, for example, has his own time-proven mixture of bonding material comprised of fiberglass, resin and filler. It comes in one-gallon containers and with sufficient catalyst to set you up in the Vette business. It's called Fiber-Weld and while it isn't cheap (no glass bonding material is), it's indispensable when working over one of Chevrolet's plastic cars.

2 No, this isn't a '78 Vette with its big wraparound rear glass hinged in hatchback style, it's a '77 with Bruno's '78-style conversion kit—and it's hinged in hatchback style. If you have a silver anniversary model and want the rear to open, Bruno can supply you accordingly. If yours is an older Vette but you like the sleek styling of the '78, then opt for the kit—and specify if you want the glass to hinge or not.

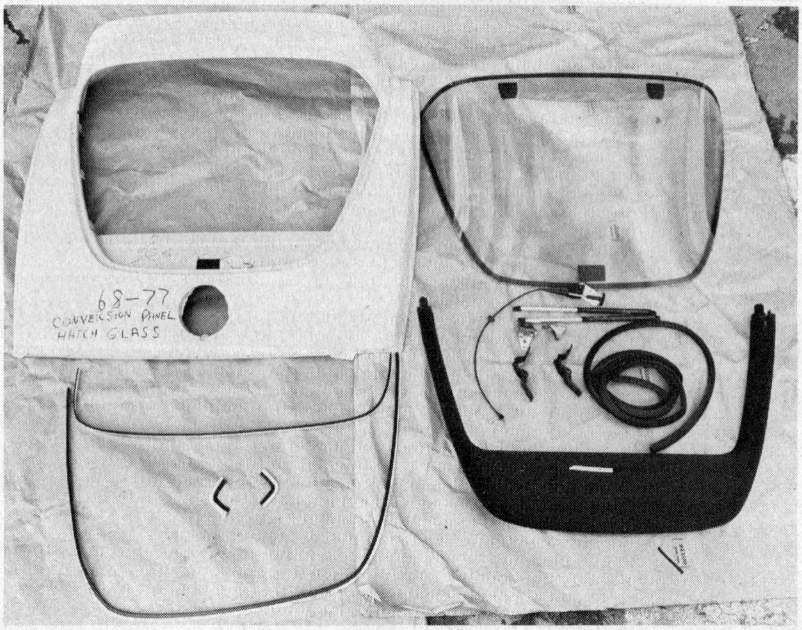

fiberglass, and as such they are easily added to the glass-bodied car. All it takes is a quantity of any of the many fillers available, resin with catalyst, and perhaps some strips of glass cloth if the item being added is heavy and strength along the joining lines is required. There are how-to's after this chapter that detail the adding-on of a hatchback roof for '68-up Vettes, and a fender-flaring job that pretty much applies to them all. Thus we needn't give you the details here of adding any of this smattering of Vette-only objects. The goodies illustrated have been picked at random just to whet your appetites, but rest assured that for every item pictured, there are a dozen more that aren't. A rundown of many firms offering Corvette aftermarket items is listed on page 47, and offer catalogs of their add-ons.

If you're a Corvette owner, or if you have Vette customers that want to one-up each other, scan the following rundown.

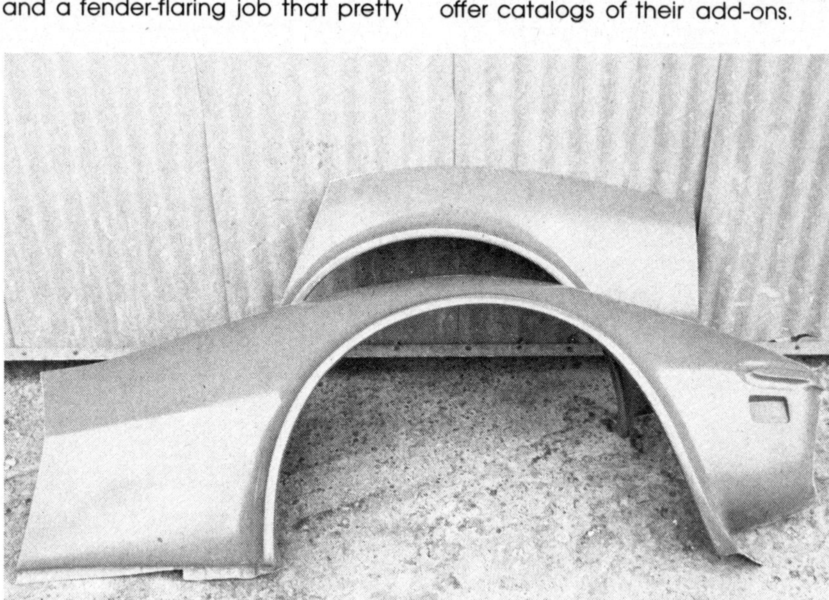

3

3 Among the many bits and pieces that J&D Corvette Automotive makes available for all '63-up Vettes are trick fender panels like these. The flared lips are on a par with the stock fenders, but the radius of the wheel opening has been increased to clear fat rubber. It's a whole lot easier to replace the stock panels with these than to get involved with all that messiness of cutting the stock opening bigger then forming new lips. They also have front and rear-end "clips", spoilers, air dams, "hard" glass nose and tail panels to replace the "soft" factory pieces of the later cars, and anything else you and/or your customer might desire.

4 Never to be overlooked are swap meets when scouting around for body, trim and mechanical pieces for older Corvettes. The ordinary car meet, or a Chevy-only gathering if you can find one, always turns up new-old-stock as well as aftermarket Vette pieces. That hood leaning against the truck, for example, was "boss" to someone once, is now a castoff but may be just what you're looking for. Good buys can often be had on complete cars, too. The wise customizer with shop down-time might buy that '60 on the trailer, either restore it or rebuild it into a custom, and turn it over for a neat profit.

4

The Corvette World

1

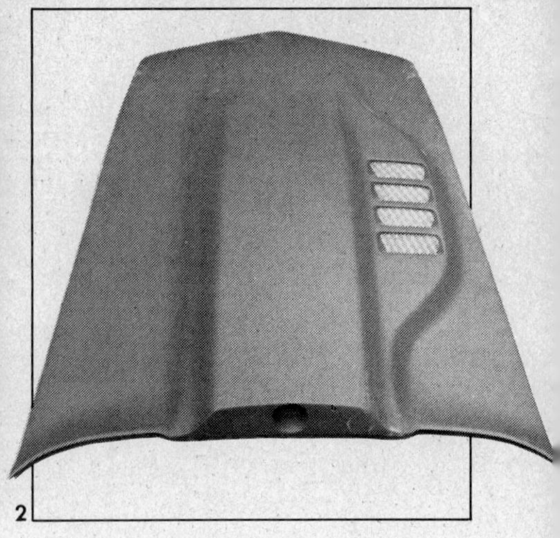
2

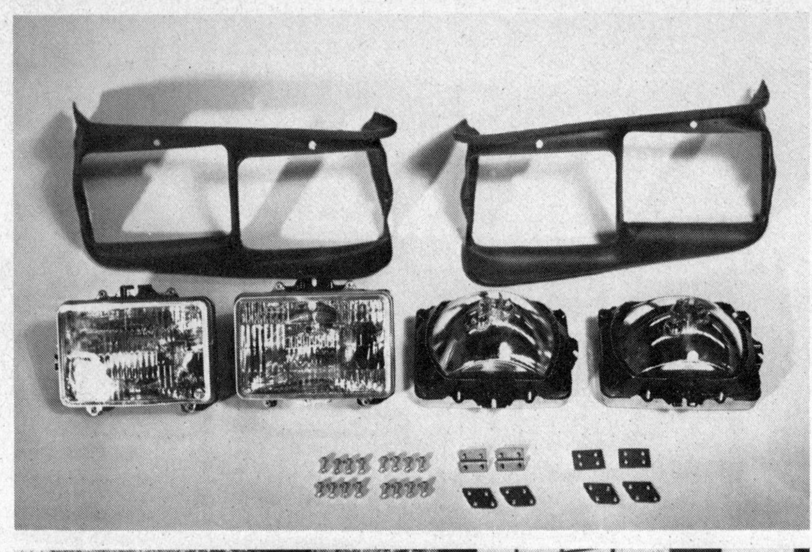

3

1 Eckler's Corvette World is no small operation; just look at the tooling and special fixtures required to meet the demands of Vette customizers. Obviously no backyard operation, it produces most any custom or replacement glass part for any year car you can think of, from a small fender patch, to a raft of rear and front body parts, complete end "clips", one-piece flip-up front ends, doors . . . you name it. Their catalog lists more parts than we can, so check them out in our service listing on page 47.

2 Attention, all you huffer fans. Here's what you've been waiting for. Ever try and squeeze a turbocharger onto a Vette so the hood will shut? Bruno has the answer to this hassle in the form of his brand-new Turbo Hood. The screened-over intakes on the right of the assymetrically shaped piece lets air get to the turbo, and the hump will handily house the related plumbing. Then to keep from punching a dashboard hole or hanging a dial anywhere in the Vette's tight confines that space permits, you mount the boost pressure gauge in the baffle at the rear. That's real class—in glass!

3 The Vette Shop can help you put square pegs . . . errr, headlights . . . into round holes. All you need to square up the lights in your '68-up Corvette is this adaptor kit. Everything to make the swap is included, but you must have your own Phillips screwdriver, some electrical tape, a hacksaw blade and a pencil. It is an admittedly small change to your car's personality, especially since the lamp lids are only open when it's dark. But you'll still appreciate being square when everyone else is going 'round together!

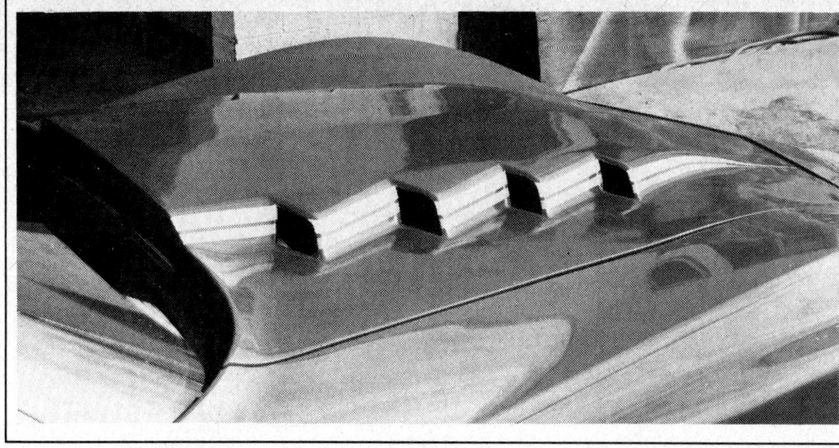

4

5

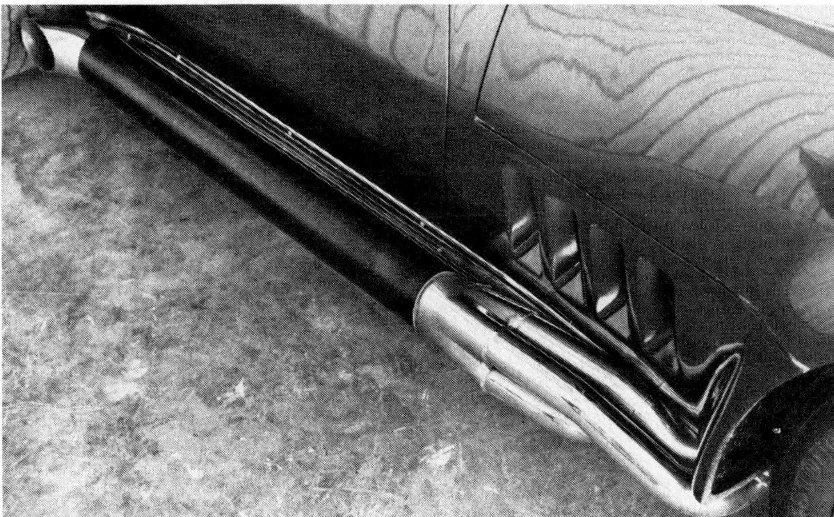

6

7

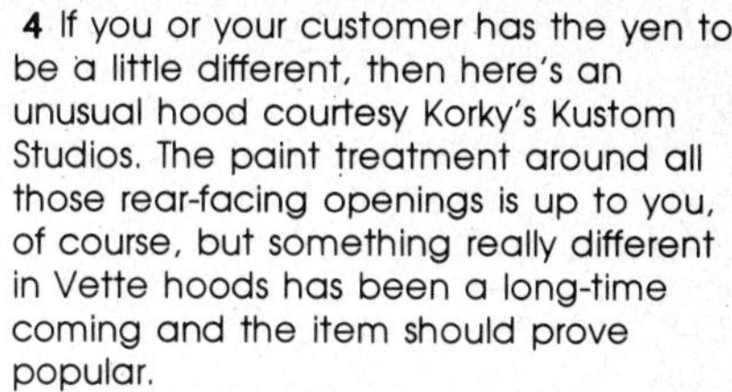

4 If you or your customer has the yen to be a little different, then here's an unusual hood courtesy Korky's Kustom Studios. The paint treatment around all those rear-facing openings is up to you, of course, but something really different in Vette hoods has been a long-time coming and the item should prove popular.

5 J&D Corvette Automotive has repair and customizing kits galore for Vettes of all ages. This is their stock replacement front end for plastic cars with bashed noses, and it's interesting that all the pieces, including inner wheel wells, are assembled which will save you a lot of alignment time. While the mating edges of abutting panels are solidly bonded on the back sides, the exterior seams are left as-is for you to finish. This makes the nose kit that much less expensive. Or, you can have the same type of assembly but with fenders flared to any degree. J&D's catalog lists all the combos available plus all of their other cosmetic parts.

6 Eckler's Corvette World, like most suppliers we are including here, offer far more goodies than cosmetic parts; performance equipment is also a big piece of the Vette pie. While not many items can be both functional and eye-appealing, these snakey Eckler headers can. They exude performance and a smarter profile, in one fell swoop

7 Just thought we'd toss something to all of you painters. So there's no such thing as a Vette paint kit? Maybe not in a box, but assemble your own with assorted colors, add some imagination, then get your spray equipment ready. The sky's the limit on Corvette painting. Maybe star-spangled inspiration isn't for you, but even a nearly stock-bodied plastic car will take on new distinction and and a different personality with a painting treatment of your own invention. And remember that the best thing about a paint-only customizing job is that your (or your customer's) Vette can be easily made box-stock again for that future collector or restorer/investor by re-applying the original color.

1

1 Can un-customizing be mentioned in a book on customizing? You bet—in the world of Corvettes, at least. With so many cosmetically-altered Vettes running around, restorers and others interested in the marque as investment potential are hard-pressed to find box-stock older models. They are beginning to find that it's less expensive to pick up one that's been modified and turn it back to stock, than to find an unaltered one. For example, try and find a '61 without fender flares, triple taillights, or whatever, like the one illustrated. Aftermarket suppliers to the rescue! Most leading Vette suppliers make, or at least handle, originally shaped body panels, trim pieces, and so forth. Check through any of their catalogs or parts lists.

2

2 Bruno's Corvette Repairs' body panels' list is too long to even begin describing it here But for starters, he can produce not only every panel for every Vette from '53 through '78, but he can provide just a part of any piece. For example, if you've tweaked the lower edge of a fender—or a hood corner, or an edge of a decklid—he'll lay-up for you just that portion with enough surrounding area to allow you to patch it right in. Bruno's custom pieces' list is extensive, too. Here one of his GrandAm-styled front end "clips" is being put on a customer's car, together with ultra-swoopy rear fenders.

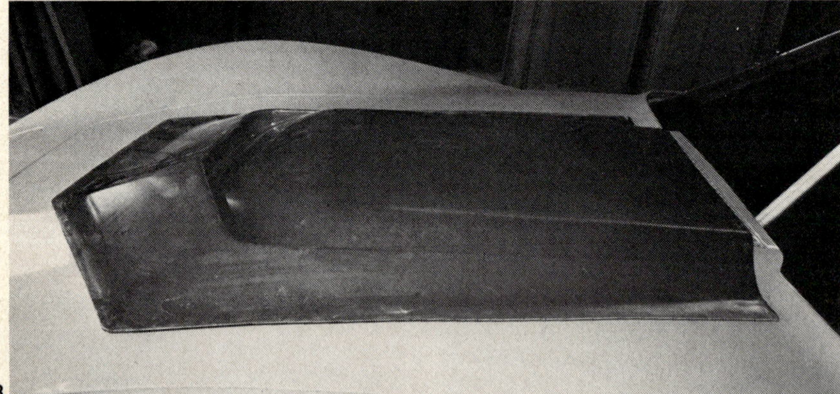
3

3 Let's say you had a carb fire, or let's say you just groove on the L-88 hood blister; either way J&D Corvette Automotive can help. Buy their entire hood, or opt for just the blister itself and graft it in place. If you want J&D to do the grafting for you, you might just as well get the entire hood—pricing is about the same. But if you can handle glass work, do it yourself and save.

4

5

4 Korky's Kustom Studios is without peer in the louvered rear window panel world, supplying them not only for any sporty-type car you care to name through new car agencies as well as custom and speed shops, but he's cornered the market at the Detroit factory-installed level as well. But he's also into Vettes, and in a big way. None other than Korky himself poses with the pieces for one of his kits which will fit Vettes from '68 and up; swoopy front and rear fenders as well as an interesting tail panel that incorporates Monza taillights. Installed on the car around which the pieces are arrayed is Korky's louver panel.

5 Hey, all you lucky 25th Anniversary Corvette owners, here's something guaranteed to put down all the other '78 Vette owners in your neighborhood. Korky's Kustom Studios has one of his louvered rear window panels all set to go on your silver car. They'll keep the sun from blazing through that big, wraparound rear glass, but hamper visibility scarcely one whit.

Corvette Body Parts Suppliers

The following listing includes manufacturers and suppliers of Corvette body parts; stock replacement panels as well as custom items. We cannot guarantee its completeness and any possible omissions are purely inadvertent. Most of the firms and services listed offer more than the fiberglass pieces that make up an any-year Vette's exterior, so if your needs are for other than cosmetic panels—engine performance equipment, heavy duty drivetrain products, special chassis parts, etc.,—write for a catalog. Most firms today charge a nominal amount for their literature, nearly always refundable with the first order, but we include only those here where the catalog and brochure prices are known. Your Chevrolet dealer parts books and factory supplements also list special body parts for the later Corvettes, so give your local Chevy agency a try, too.

ACME FABRICATING CO.
2409 17th Street
San Francisco, CA. 94110
Catalog: $1.00

AL KNOCH'S INTERIORS
32848 Manor Park Drive
Garden City, MI. 48135

AUTOMOTIVE PLASTICS CO.
P.O. Box 408, Pleasant Ave.
Geneva, OH. 44041

BERGER CHEVROLET
2525 28th Street, S.E.
Grand Rapids, MI. 49508
Catalog: $2.00

BRUNO'S CORVETTE REPAIR
11055 Ventura Blvd.
Studio City, CA. 91604

CORVETTE AMERICA
P.O. Box 427
Boalsburg, PA. 16827
Catalog: $4.00

CORVETTE BODY & SERVICE
2302 Harrison Street
San Francisco, CA. 94110

CORVETTE SPECIALIST
13704 Hanford-Armona Rd.
Hanford, CA.

CORVETTE SPECIALISTS OF PASADENA
1835 E. Walnut Street
Pasadena, CA. 91107

DICK GULDSTRAND ENTERPRISES
11924 W. Jefferson Blvd.
Culver City, CA. 90230
Catalog: $2.00

ECKLER'S CORVETTE PARTS
P. O. Box 5637
Titusville, FL. 32780
Catalog: $2.00

FASTGLAS
56 Park Avenue
Penndel, PA. 19047
Catalog: $2.00

FERRIGAN & O'BRIEN, INC.
556 Cortland Street
Albany, NY 12208
Brochures: Free

IVAN BAILEY
715 Doan Drive
Burbank, CA. 91506

J&D CORVETTE AUTOMOTIVE
13512 Lakewood Blvd.
Bellflower, CA. 90706

JOHN GREENWOOD SALES, INC.
2700 Princeton Street
Dearborn Heights, MI. 48127

J's AUTO RANCH
1842 Kutztown Road
Reading, PA. 19604
Catalog: $0.75

KORKY'S KUSTOM STUDIOS
15430 Cabrito St.
Van Nuys, CA.

McCARTY'S FIBERGLASS FABRICATORS
Rt. 1
Mineral City, OH. 44656
Catalog: $2.00

"MCR" CORVETTES
5115 So. Industrial Road
Las Vegas, NE.

MEL'S AUTO PARTS
871 Garden Hwy.
Yuba City, CA. 95991

MICHAELIS CORVETTE SUPPLIES, INC.
Rt. 1, 424 East
Napoleon, OH. 43545
Catalog: $2.95

MOTION PERFORMANCE PARTS, INC.
598 Sunrise Hwy.
Baldwin, NY 11510
Catalog: $2.00

R. C. CORVETTE PARTS
1439 So. La Brea Ave.
Los Angeles, CA. 90019

REDAN INTERIOR SPECIALISTS
147 Franklin Ave.
Hartford, CO.

SONNY'S CORVETTES
25325 So. Normandie Ave.
Harbor City, CA. 90701

THE VETTE SHOP, INC.
23859 Telegraph Road
Southfield, MI. 48075
Catalog: $2.00

THOMPSON TOPS AND PERFORMANCE
P. O. Box 2002
Florence, AL. 35630

THUMPER SPOILERS
P. O. Box 6028
Torrance, CA. 90504

HOW-TO: CORVETTE HATCHBACK KIT

Because fiberglass is a relatively easy base material to work with in forming shapes, and because the molds that constitute the tooling are much less expensive than steel stamping dies, there are many aftermarket firms offering a wealth of glass add-ons. From items as small as rear mirror fairings, to complete automobile bodies, the glass add-ons list is as diversified as it is long. Fender flares, spoilers, air dams, scoops—they're available in a potpourri of designs and sizes, and because of the nature of glass, they can be as easily added to a glass-based Corvette as they can to a metal car. Besides the trick aftermarket pieces, "clips" or stock-shaped replacement panels are available for Corvettes from many sources, as are replacement panels of exotic shapes.

One of the major Corvette replacement panel builders is Southern California's Bruno's Corvette. He stocks—or can quickly produce from molds on hand—literally any body part for any year Corvette, as well as front and rear body sections of other-than-stock configuration. One of his newer kits turns the 1968-up Vette into a hatchback. Bruno will either sell you the kit and you can install it yourself, or you can drive your car into the busy establishment and have them do it for you. The job, while straightforward as far as adding-on glass pieces is concerned, is not really for the faint-of-heart. Nearly all of the stock body from the doors aft, except portions of the fenders, have to be removed and discarded. Slicing so much material from a car as expensive as a Corvette may give you qualms—either about your own car or somebody else's—but the instructions that accompany the kit are concise, accurate, and include everything needed to complete the job less the resin, glass cloth and epoxy-based filler—things any body shop that does Corvette work has on hand.

The kit components include the hatchback panel, the tail section and the inner floor piece, all of glass. Also in the kit is the liftgate with all hardware and glass (from a Datsun Z-car), taillights, rubber moldings, upholstery, and new gas tank brackets for the supplied Camaro tank to go under the body floor.

1 The top section of this Corvette from the doors back, the deck area, sail panels and rear body section will have to be cut off to allow installation of Bruno's Corvette hatchback.

2 Not pieces from our subject car, this will still give an idea of how much of the body has to be cut off. Cuts are made at factory seams, and the kit instructions locate them.

3 New tailpiece is essentially of stock Vette configuration except that upper edge is angled to conform to slope of new roof panel. Not shown here is the included inner pan.

4 Original top is removed from integral steel hoop at rear door line, then all traces of factory cements and glass are ground off for perfect adhesion of the new roof panel.

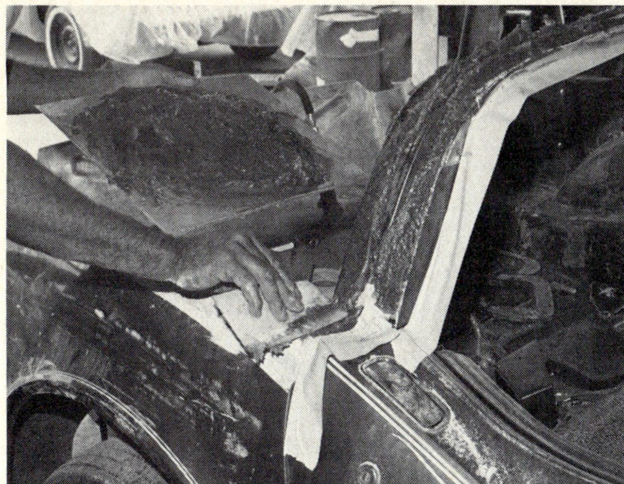

5 Bruno uses—and can supply the bodyman with—his own formulated filler, known as Fiber Weld. It's a two-part mixture like most epoxy-based fillers, here being applied.

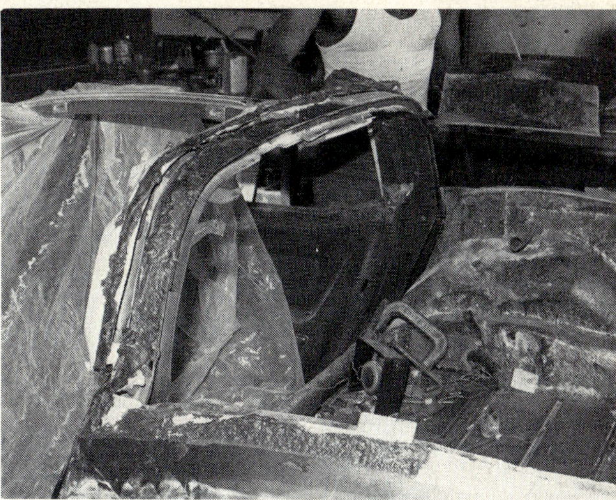

6 Plastic sheet covers cockpit to keep out as much ground fiberglass as possible. Filler is also applied at front lower edge of roof panel, but is not yet put on further back.

7 Clamps secure front edge of roof panel to steel hoop while the filler material sets up. Trial fits and checks are, of course, necessary first, as with any add-on product.

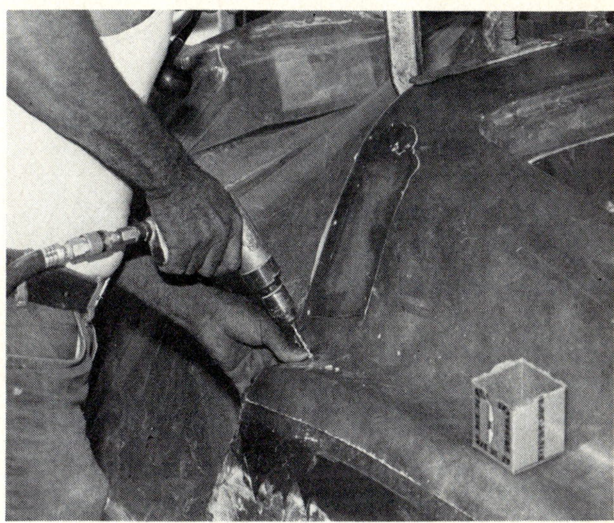

8 Top panel overlaps cut line on body where old roof was removed. Holes for temporary sheetmetal screws are drilled to secure mating pieces while the filler cures.

9 All glass surfaces must, of course, be ground down to clean base material. When adhesive filler has cured, screws are removed and more filler wiped over the line.

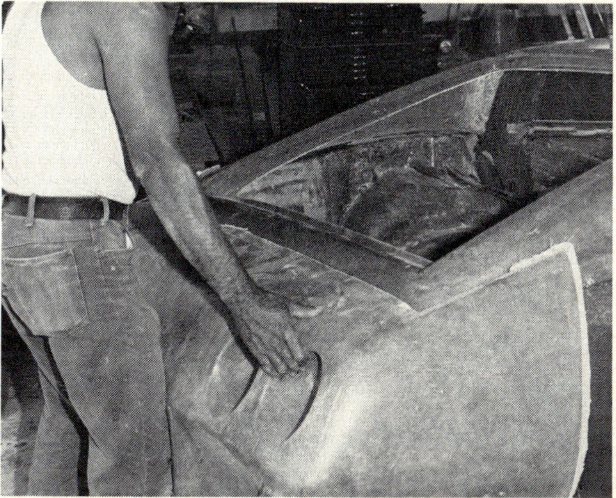

10 Tail panel is a single piece, is purposely made to heavily overlap mating surfaces. Precisely mated edges cannot be produced due to a great variation in the body parts.

11 Pneumatic drill motor with small circular saw blade zips off excess glass. Final transition of surfaces through this area must be smooth, so a butt-edge is preferred.

12 Scrap metal is metal screwed to hold pieces during final alignment, and while inner floor section is being added. Gas filler door must be added, cut from original body panel.

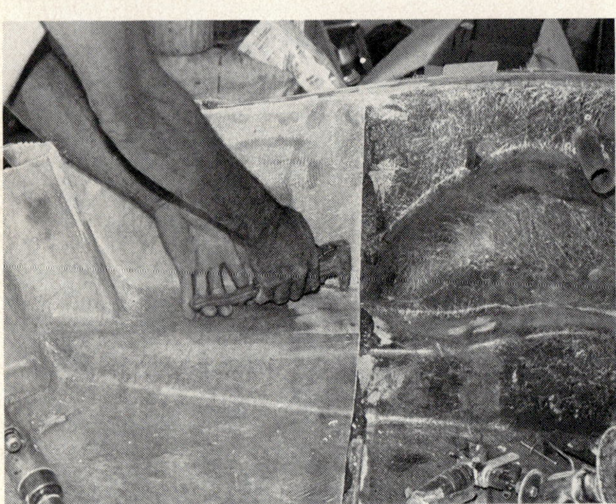

13 Pop-rivets make quick work when adding new inner floor pan. This means use of a Camaro fuel tank; together with hangers supplied to mount it where stock spare is.

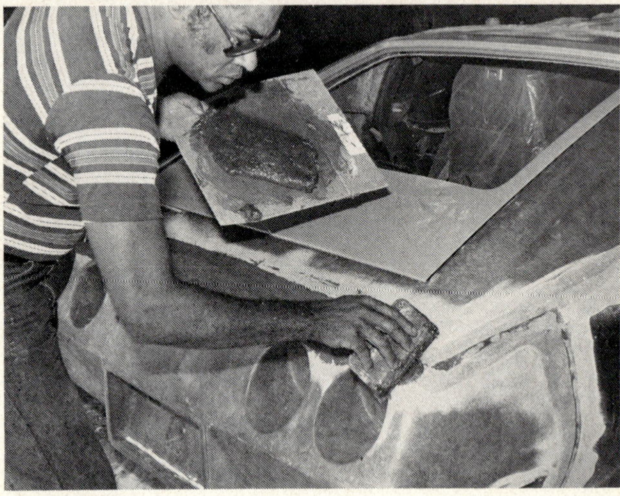

14 Lapped filler, when cured, is ground smoothly down, more added to fill any voids. Datsun hatch, kit-supplied, is in place here to assure a uniform fit around opening.

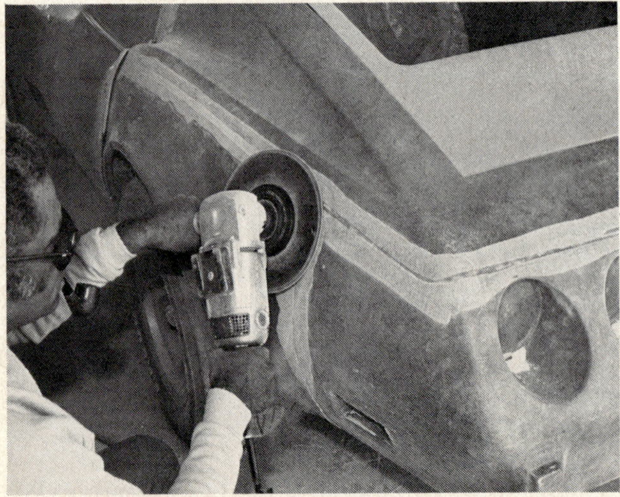

15 Another grinding-down of filler is needed before layer of resin-impregnated glass cloth is added to complete union of the parts. Final result will be stronger than original.

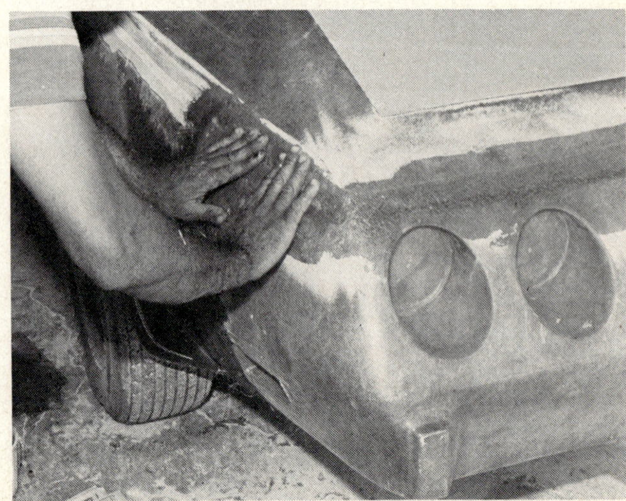

16 Strips of glass cloth are soaked in resin, placed singly along all seams. Curing is fairly rapid, then must again be ground smooth. Note that taillight holes have to be cut

17 A final grinding now, then bondo or similar filler wiped over all areas to fill any imperfections. The "mess" of the Corvette a few steps back is beginning to shape itself up.

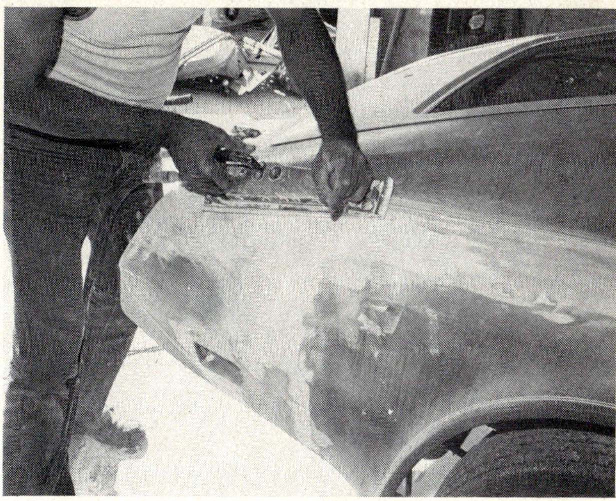

18 Pneumatic file makes quick work of Bondo, but must be used with caution to avoid cutting too deeply. Next step, of course, is careful sanding of whole car, then priming.

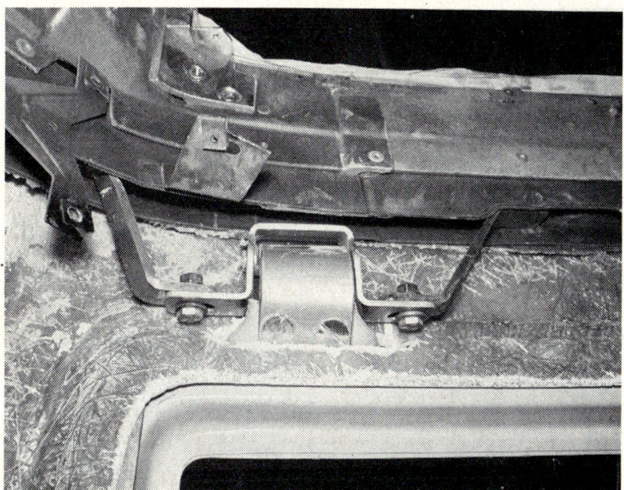

19 Detail of hatch hinging, utilizing part of the Datsun mechanism with other pieces supplied by Bruno in the kit. Bruno's kit runs about $1500; installation is worth $1000.

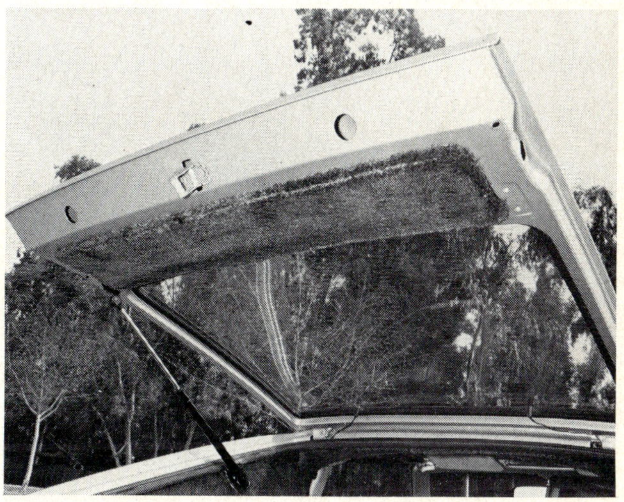

20 Completed kit, hatch raised. Telescoping strut, latch, glass, etc., are all supplied and all are Datsun. Kit includes all upholstery, including trunk-area mat.

21 Fuel filler was cut from original body, mounted thus. The Corvette emblem on hatch also is taken from original car. Kit does much to change Corvette's appearance.

22 Some 12 cu. ft. of additional interior space is added by kit and replacement of fuel tank, which now occupies former spare tire location. Car is shown in color on our cover.

HOW-TO: FIBERGLASS FENDER FLARES

1 The first step in any add-on kit project is to trial-fit the pieces. Ultra-wide flares are aligned, perimeter marked, then flares are removed and paint ground to bare fiberglass.

2 Fender flares themselves don't require removal of stock body material, but this Corvette is slated for super-wide tires, so wheel openings are enlarged using an air chisel.

3 There is sufficient wheel rebound and steering space now. Inherent glass strength means enlarged wheel opening can be left raw-cut. Skirt cut-out will duct air through scoop.

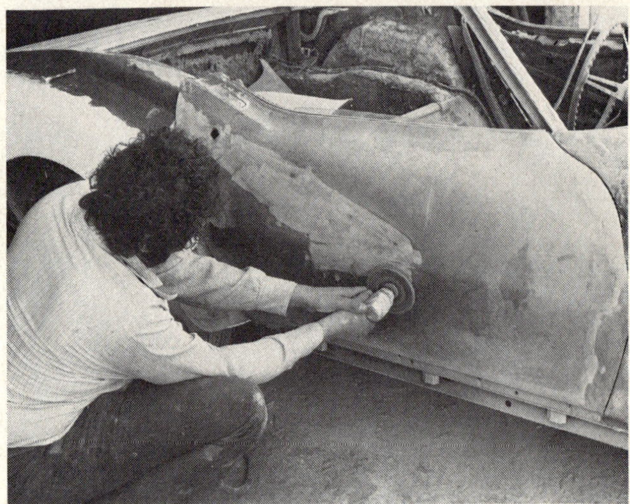

4 Bonding with fillers or resin-impregnated cloth to glass needs removal of all traces of paint and primer, down to the base material so adhesion is complete and strong.

5 Flares are repositioned and attached temporarily with 1½-in. metal screws. Tabs are temporary, but are molded to Chevrolet-made racing flares to aid positioning.

6 J&D Corvette plastic filler is mixed with its catalyst and spread liberally along flare-mating line. The flares are mounted one at a time since the filler sets up rather quickly.

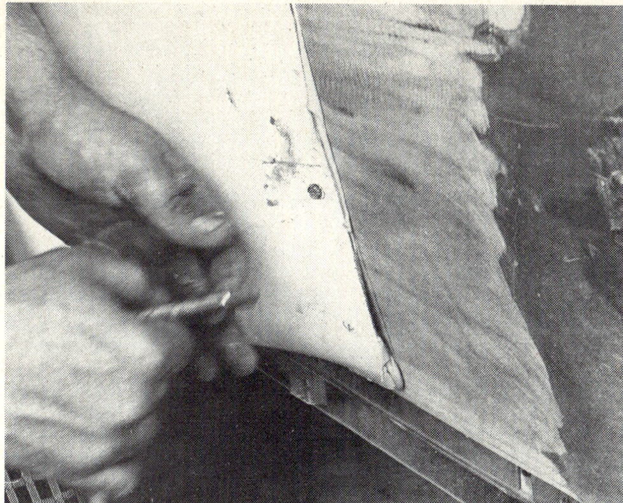

7 The flare is pressed against the filler and sheetmetal screws driven in to pull the add-on tightly against the panel. Screws remain only until filler hardens, then are removed.

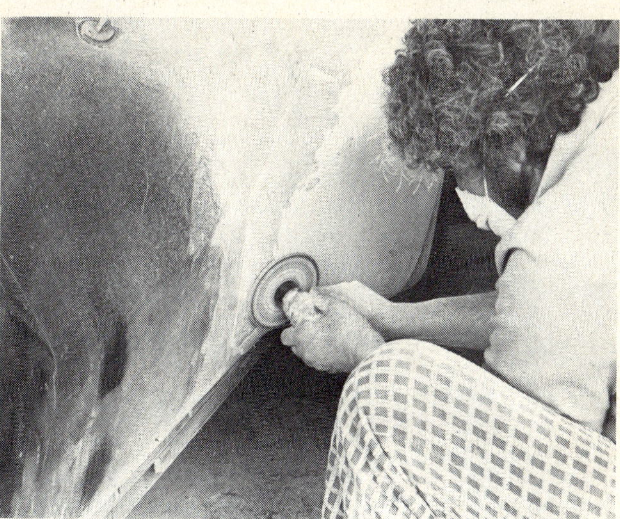

8 An area several inches to either side of mating line is ground down to remove excess filler and to assure a clean surface for the further adding of resin and filler material.

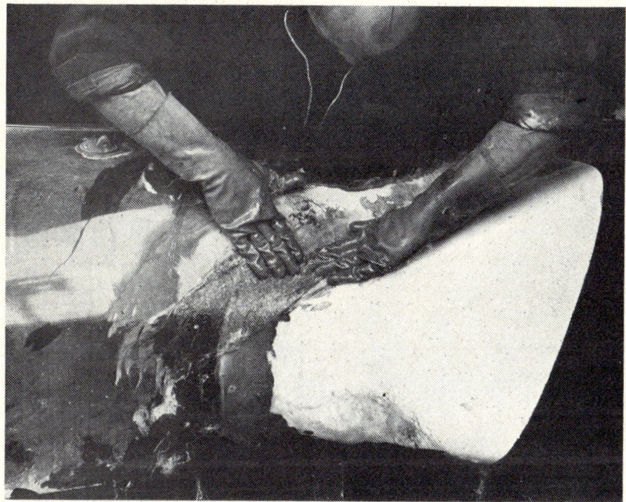

9 With hard driving in this Corvette's future, strength is needed to keep the fairly heavy flares from pulling away from the body, so resin-soaked cloth is laid over the joints.

10 After the fiberglass has cured, it is ground smooth then additional filler is wiped on to fill imperfections and to add a fillet. When cured, it will be filed and jitterbugged.

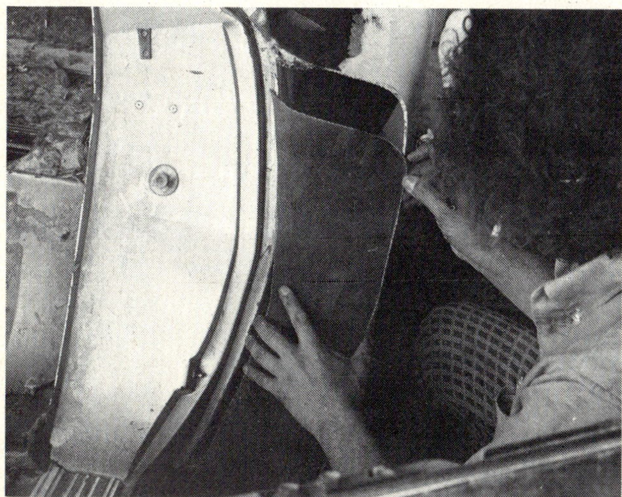

11 New rear flares extend forward onto doors, so after each is installed, door-opening line is cut with an air chisel. Then, panels are cut from fiberglass to fill openings.

12 From here on out, it's all up to the painter, but the visual change in the Corvette that flares make is evident. Wheelwell voids will be better filled with fatter tires.

Carving a Custom

Saw, file and sand a shape to your own design.

by Eric Rickman

The long established procedure in creative body design is a method pioneered in Detroit. From preliminary sketches, artist renderings, and scaled blueprints a three-demensional clay mockup of the design is made to get an overall picture of the newly created shape in its completed form. The primary advantage of this process was the flexibility of the clay model, if the shape needed refinements, just add more clay and sculpt a new shape incorperating the desired changes. From final clay model to the first sheetmetal prototype involved a lot of hard work by very skilled metal workers.

The customizer, working with the existing vehicle doesn't have the advantage of the quick design change of the clay mockup, he must fabricate his new design in sheetmetal, lead and Bondo right on the car. This is time-consuming, hard work and offers no design change flexibility, if you don't like the results the whole thing must be redone; in metal this is both expensive and difficult.

With the advent of the age of plastics a new process has become available that combines the flexibility of clay and the ease of fiberglass construction.

We are indebted to Gordon Saunders of G.S.A.I.D. for a short course in the use of this method to translate a new design shape into the three-dimensional finished product.

By using polyurethane foam in place of clay we can literally hand carve the desired design into the finished shape with a minimum of labor.

Urethane 190 is available from the C.P.R. Upjohn Co., (the drug manufacturers) in 4x8x2-ft. billets for approximately $143. Reject foam and scrap is available for less and can be cemented together to achieve the size block needed for your particular job. Upjohn manufactures urethane foam in various densities, however the 2-lb. density has been found to be ideal for this type of work. This material can be sawed, filed, sanded, and carved much like a block of soft wood, like balsa, only much more easily, and can be worked on the bench or glued directly on a vehicle and carved into the desired shape in place.

Foam has all the flexibility of clay, in that if you goof, or don't like a particular line or shape, more foam can be cemented or foamed over the area and you can start over. The medium is completely flexible permitting in-progress design changes with a minimum of labor and expense.

FOAM CHOICES

There are two ways to go with this method of foam sculpting; the one-shot, on-the-car custom designing, or you may wish to duplicate the shape in a number of identical parts. In either case, the sculpting can be done on the car to be sure the overall concept is exactly what you want before the design is fixed.

The difference in the two methods is that in the one-shot process the foam is permanently affixed to the car and becomes a part of the finished job, while in the duplicate parts method, the foam is affixed to the car temporarily while sculpting, then removed and used as a

1

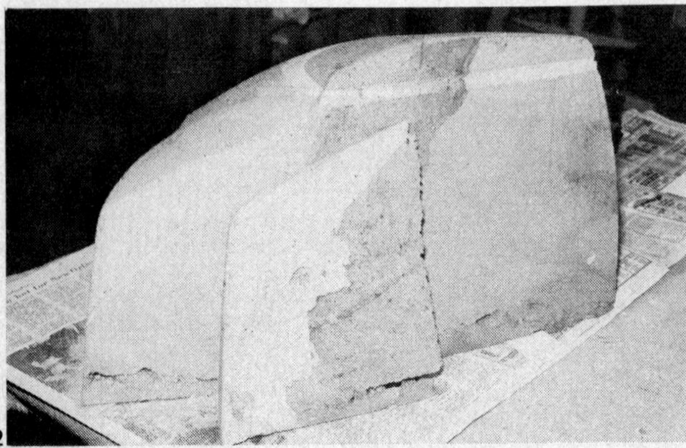

2

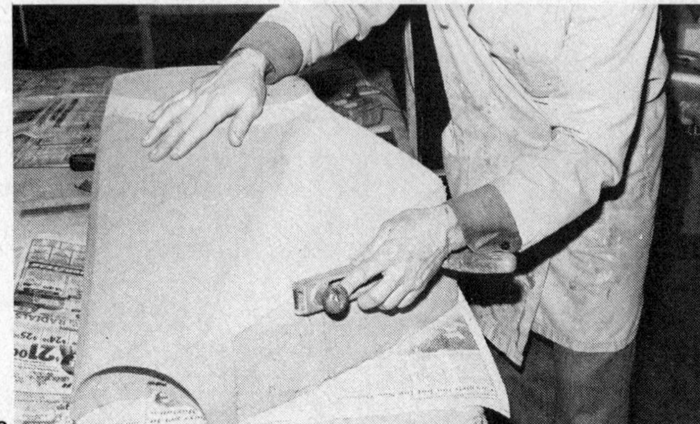

3

plug for a fiberglass mold from which duplicate parts can be produced.

The secret in cementing foam pieces together, or cementing the billet to the car, is to use a two-part foaming solution that when mixed reacts chemically to foam up and set into a rigid, cellular, non-porous mass of almost the same density as that of the Urethane 190. This two-part foaming material is available from most boat shops, as it is used to fill voids for a boat's flotation. Another advantage of the foaming material is that it can be poured over goofs and low areas to build them back up so you can reshape the surface. Just use small cardboard dams taped in place to confine the foam for a short time until it hardens.

To cement a foam block to a car permanently, rough out one side of the block to fit the side of the car, pour in the foam mix and press the billet against the side of the car until the foam sets. The fit needn't be too precise as the foam will fill all the voids very nicely, and it is very hard to get off. If you only want to attach the foam billet temporarily, use a few dabs of Bondo to hold the foam in place; it can be cut away later and the Bondo sanded off the car.

The use of liquid foam as a cement results in a homogeneous mass that will sand and file evenly. Any other cement would produce a seam of a different density that would give trouble when trying to achieve a smooth surface over the cemented joint.

With the foam block in place, either on the car or on the bench, the shaping process is begun by sawing away the excess material to achieve a rough outline. A disc sander can be used to work the foam down to the approximate shape, then final shaping is done with cheesegrater files and sandpaper, both flat and curved. Flat sanding boards are made from small pieces of plywood with wood-block handles. The sandpaper, in this case #36 grit floor-sanding paper, is attached to the boards with 3M HiTack spray cement. The #36 grit paper is used for rough shaping because its coarse open grit doesn't load up so readily. Progressively finer paper is used as you near the final shape and surface finish. To get into curves and fillets sandpaper is affixed to various diameter cardboard tubes.

At this point it should be kept in mind that the foam form must be made slightly smaller than the desired finished dimension, as it will be covered with glass cloth if it is to be used as a mold plug, and with both cloth and matte if it is to be left on the car.

With the newly created form

1. Rigid polyurethane foam can be cut very easily with a fine tooth saw. It can also be carved with a knife.

2. Stratification marks reveal the areas where two pieces of block were cemented together with liquid foam. Note foam buildup on left side.

3. Stanley cheesegrater files can be used to good advantage to shape the foam block with ease.

4. Rough shaping can be done with a disc sander, then final shaping is done with sanding boards to which sandpaper has been cemented.

5. Open cell surface of foam is now sealed with a filler paste, (see text). This fills foam cells and any air bubbles that may be present.

6. After paste is squeegeed into the surface, excess is struck off with a straight edge to retain contours.

7. Any of the resin/catalyst steps can be accelerated by heating the surface with a heat gun if the day is cool enough to slow the process.

8. Cut two pieces of glass cloth big enough to cover the foam shape. Allow extra for overlap if shape is on a car body, foam will have to be molded to the car body.

9. A liberal coat of resin/catalyst mix is applied over the sealed foam surface. We are working with part of the foam shape to demonstrate.

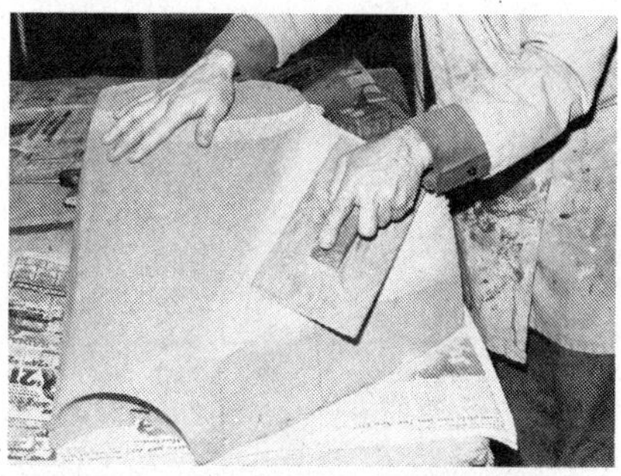
4

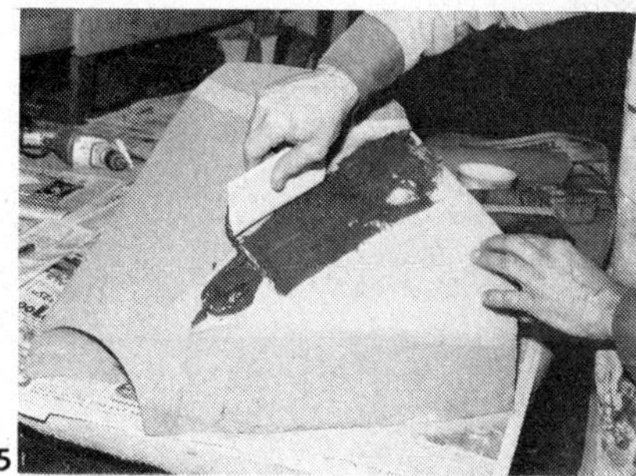
5

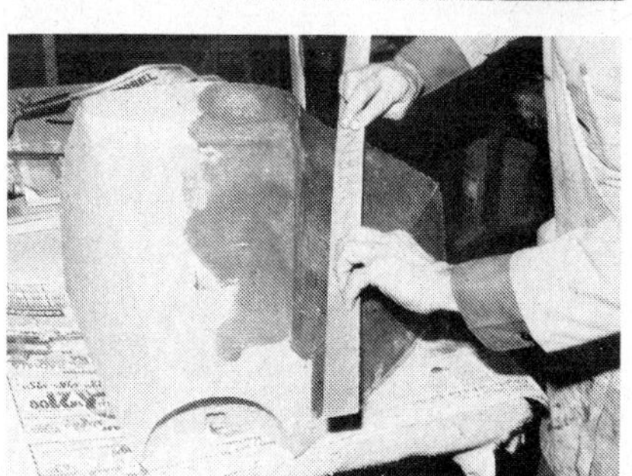
6

7

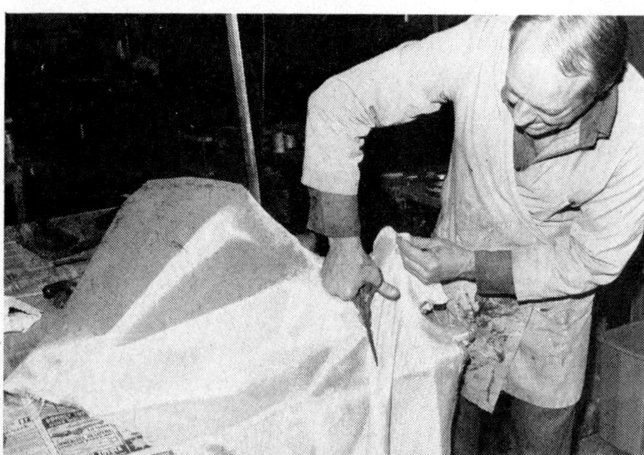
8

9

completely shaped and sanded smooth on one side of the car, it will suddenly dawn that this same shape will have to be mirrored on the other side of the car.

Detroit does it with mirrors; only half a clay mockup is made against a mirror and the reflection provides a simulation of the entire vehicle that can be seen from any viewpoint on one side of the car. This isn't going to be that easy.

A baseline must be established down the length of the new shape, this baseline is located from fixed reference points on the car body, such as the top of the fenderwell openings, an exposed belt molding, or chrome trim. The lines are drawn down the length of the foam block so they will be equidistant from, and parallel to, the fixed reference points on both sides of the car. Next, stations must be laid out at equal distances along the length of the baseline from front to back; the more complex the shape, the closer the stations will have to be. A template of the new shape must be made at each station of the finished side. These templates are used as guides when shaping the foam on the opposite side of the car to make sure the finished job will be symmetrical.

When working with the foam affixed to the bench top, templates are often made directly from blueprints or full scale drawings to be sure the foam plug matches the drawings. The templates are then reversed to reproduce the other side of the car.

When the foam is sanded down to its final shape it must be sealed to fill all the tiny cellular voids and air bubbles. A semi-liquid paste filler is made by mixing resin with either phenolic bubbles or glass beads and a filler called "Cabo-sil." The phenolic bubbles and glass beads are so tiny the material is like flour, and it thickens the resin into a paste that is only slightly thinner than Bondo.

KNOW YOUR SURFACE

This thixotropic paste with a catalist is then spread evenly over the foam shape and squeegeed into the open porous surface and any slight irregularities. Excess material is then scraped off with a straight edge. Remember, you already have the desired shape, all you are doing is sealing the foam surface. After the material has hardened the surface can be sanded lightly paying attention to the

1

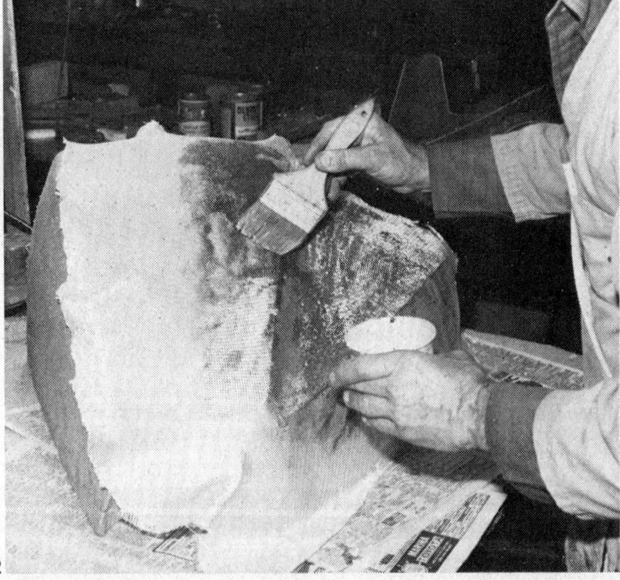
2

3

4

edges. The use of resin in this process precludes the use of Styrafoam as it will dissolve in resin.

Now is the time to remove the foam shape from the car if you are going the plug mold route; the finishing process will be the same in either case. If you are leaving the foam on the car you will just blend the glass cloth over the edge of the foam shape and on over to the car's surface to create a smooth, uninterrupted finished surface across the juncture.

Cut sufficient fiberglass cloth to cover the foam shape with two layers of cloth. A heavy coat of clear resin and cataylist is applied over the foam and a layer of glass cloth is spread over the wet resin surface. Using a small brush, additional resin and cataylist mix is spread over the surface of the glass cloth being sure the cloth is completely saturated and worked down firmly against the surface of the foam shape and there are no trapped air bubbles under the cloth.

After the first layer of glass cloth has hardened, another coating of filler paste is squeegee'd over the cloth surface and struck off with a straightedge. This will fill the cloth weave pattern so it won't print through the finished surface. After setting, the cloth/paste surface is lightly sanded and another layer of cloth and resin/catalyist mix is applied as before. This too is given a coat of the paste sealer after it hardens.

If the newly formed shape is to be used as a plug from which to form a fiberglass mold, two coats of cloth and sealer will be sufficient. If the foam shape is to be left on the car a thin layer of glass matte should be applied over the two layers of cloth to build up at least an ⅛-in. layer of fiberglass over the foam core.

With everything set and hardened you now have a fiberglass skin over your foam shape. Proceed to finish this glass surface in the same manner you would the surface of any fiberglass-bodied vehicle; block sand and fill until you have a perfectly smooth surface. The final coat is a layer of "Ram-namel" a material known as a sprayable, polyester primer surfacer. The Ram-namel surface is lightly sanded and buffed to a hard glossy finish. You can now paint the surface, or apply a gelcoat and lay up a fiberglass mold over your foam plug.

5

6

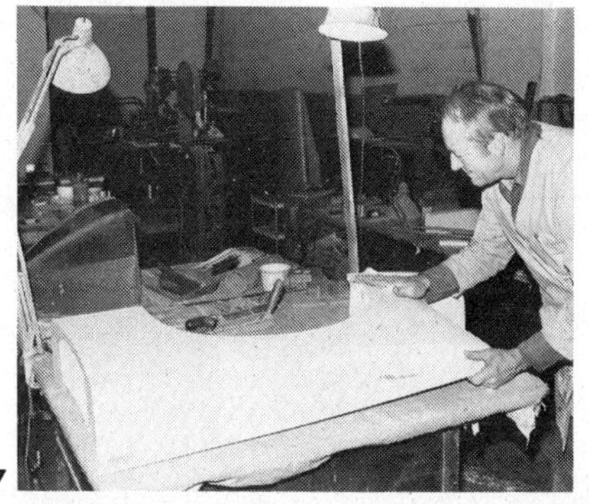
7

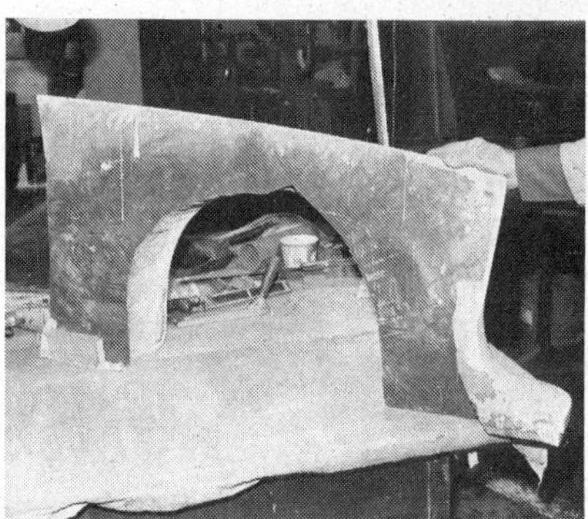
8

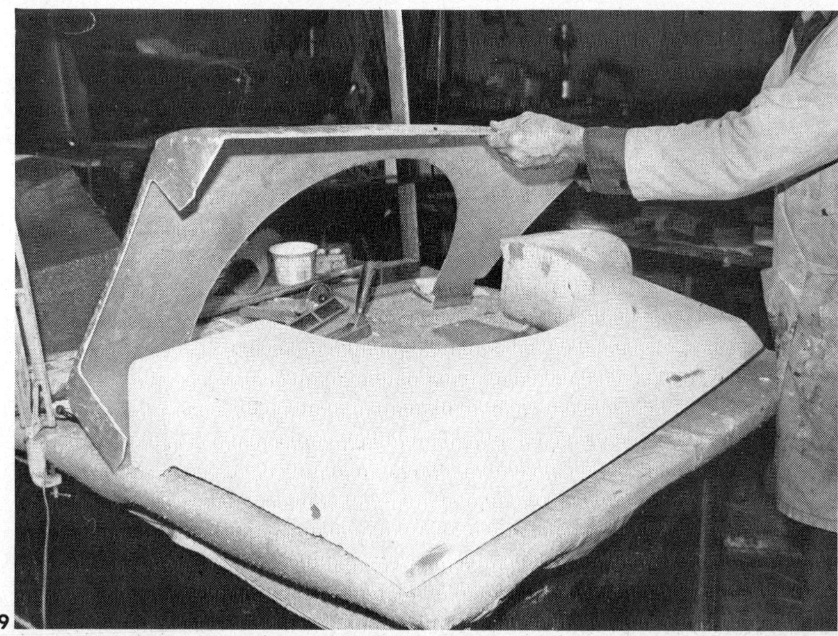
9

1. Cover the wet resin with a layer of glass cloth. Be sure it is pressed firmly into all the contours.

2. A liberal coat of resin/catalyst mix is now brushed into the glass cloth being sure to work out all the trapped air bubbles and wrinkles.

3. Another coat of filler paste is now applied after the cloth/resin has set hard. Paste fills the weave to prevent surface printing.

4. Squeegee filler into cloth surface thoroughly to fill weave, then strike excess off with straightedge. Done right, weave pattern won't show in the finished surface.

5. Sand lightly after the surfacer has set. Note cardboard tube used to back sandpaper with right radius to reach into fillet properly.

6. The second piece of glass cloth is now applied as before, and then finished with a coat of filler. If foam shape is on a car, a thin layer of glass matte will now be applied. Glassed shape is finished just as a glass bodied car surface would be, sand/fill, and sand, untill smooth.

7. Designer/craftsman Gordon Saunders gives a fender mold a final fine sanding before buffing.

8. Rear view of foam block reveals surface where block was temporarily affixed to a Ford Fiesta to sculpt a new fender shape.

9. Block could be left in place and moulded to car with glass cloth. In this case block was removed and used as a mold to reproduce fiberglass restyling kit parts. Here a fiberglass mold is lifted off foam form.

Junkland to Oakland

Building a radical custom the easy way: from a wreck

Text and Photos by Jay Storer

Building a radical custom from a late-model vehicle of any type is an expensive proposition. Unless you have an established body parts business already, you'll have to throw away a lot of perfectly good (and expensive) sheetmetal parts when you build a custom, particularly if you use fiberglass replacement parts, which is the modern way to customize. When you also consider that every custom modification limits your potential sale of the finished car to a smaller audience, economics will cause many bodymen to think twice about taking on a major customizing job for himself rather than a paying customer.

One of today's most active customizers, Carl Green, suggests that the easiest way to build a radical custom car, either as a shop advertisement or for resale at a profit, is to start with a wreck purchased at an insurance salvage pool. The major rework of a Datsun 240Z which is the subject of this story began this way. "As long as you're going to strip, cut and rework the body, you might as well start with one that is damaged," says Green. You can get into a low-mileage, late-model wreck quite cheaply, but Green offers some tips on what cars to look for at the insurance lot. Most important, look for the least amount of frame damage, especially on the unibody cars. Stay away from burned cars, rollovers, and side hits (unless above the frame). Look for a car that's been well-maintained mechanically by checking the oil, interior cleanliness, mileage, lube stickers, carpeting etc. At the salvage pool, unlike with a dealership or private party, what you see is what you get. The insurance companies don't doctor them up before the auctions.

The 240Z's shown here were both salvage jobs customized at Carl Green Enterprises. The transformation from wreck to show car for the Oakland Roadster Show took 30 days of intensive work. For the gleaming results, see page 86.

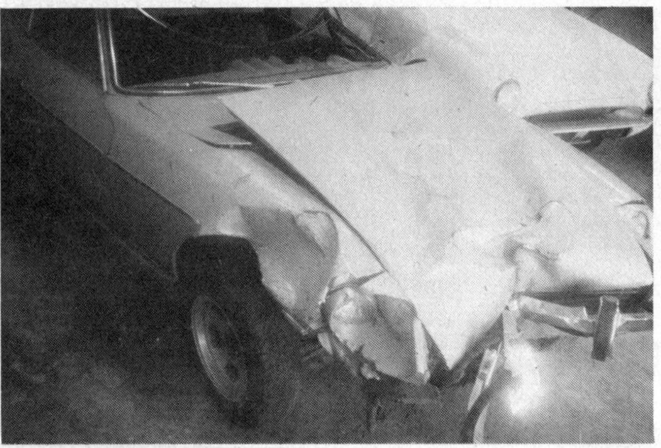

1 Knowing that he would be replacing all four fenders and customizing most else, Carl didn't hesitate to buy this 240Z at the insurance salvage yard. It was in great mechanical shape, but other bodymen passed it up as a total.

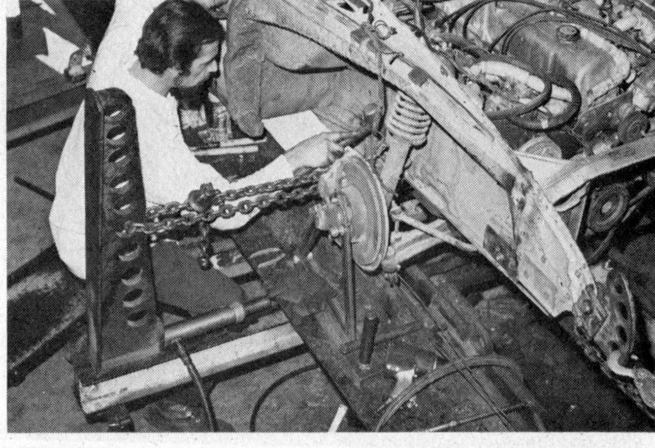

2 It takes an expert with the right equipment to straighten frames. Don't try to do some yourself to save money, you may make it worse. Tom Powers of The Frame Shop is working here with about eight tons of pressure to fix our Z's unibody.

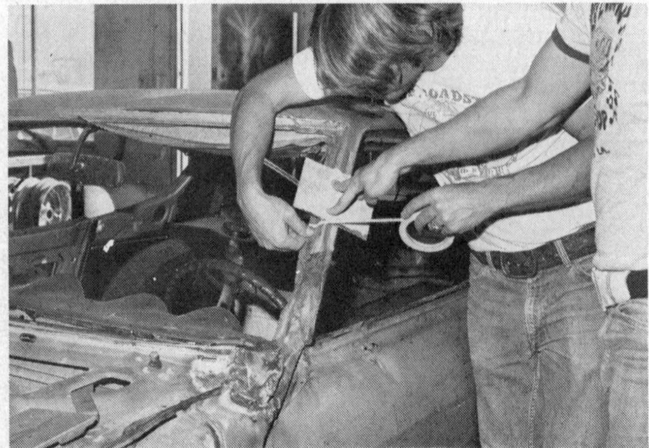

3 Customizer Carl Green uses a strip of metal sheared to the right size for alignment of the ¼-in. tape used to mark the Datsun's front posts. This damaged post will be replaced with a good one from a salvage yard. All glass is removed.

4 The secret to a good chop on a car like the Z is ¼-in. tape. The customizer runs tape, from his top and bottom marks up front back to where they just meet at the extreme rear of the car. This establishes the proper "pie section" cut.

5 An electric Sawzall is used here to cut the quarter panel along the tape lines. Go slowly, you may run into interior bracing that will snap blades. There are other ways to cut, but a saw makes the cleanest thus easiest to weld later.

6 Here you can see the point at which both cuts will end at the rear of the quarter panel. After one cut has been made, the customizer uses air chisel to separate interior and exterior sheetmetal. Heavy metal must be cut with air chisel.

7 With both front posts and both rear quarters cut, entire roof assembly slinks down as the quarter panel overlaps the top at the rear. In most chops, post alignment is handled by bending posts or adding a section to the middle of roof.

8 Before the Z went back to The Frame Shop for further frame alignment, Green explained to Jim Cook how the rear of the body was cut and the top moved forward. Jim Cook Racing sells the fenders and wing (as at left) we used.

9 Quarter panel is tacked to the roof of another Z top chop while the first goes back to The Frame Shop. Arrow indicates tacked patch to retain decklid lid gap during construction. Body was cut, bent to move entire top forward.

10 With the door glass removed, the Z's window frames are shortened and rewelded. Bending is required (A) to match the new angle of the top and post. The rear post of the frame is shortened at B, and cut at C makes new angle.

HOW-TO: RADICAL CUSTOM FROM A WRECK

11 The old quarter window graphically shows both the amount of the top chop and the new angle of the side windows, which are slanted forward to match the new angle at the rear of the body and the new rear window to be made.

12 Since the original quarter window opening is going to be filled in, the Sawzall is again employed to trim off the rear portion of the roof's dripráil. Note how smoothly the tacked quarter mates up. This car already had CGE flares.

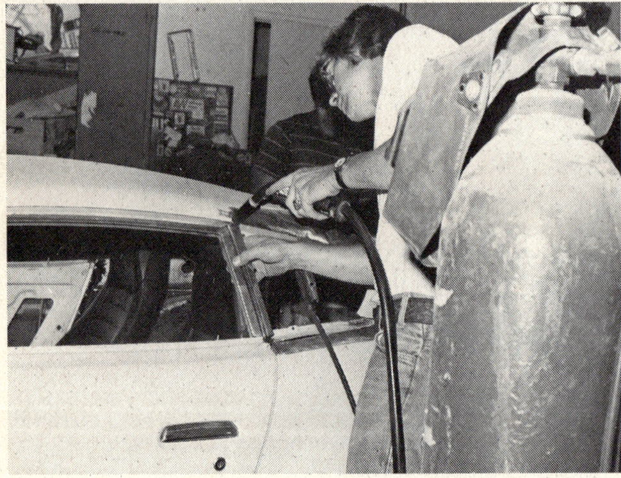

13 The front edge of the quarterglass frame is cut off and used to form a new rear jamb for the doorglass frame, for a good fit of the weatherstripping. Lincoln wirewelder is invaluable in this kind of work, allowing one-hand spot-welding.

14 There are hundreds of steps and many hours of work we can't possibly show in one article. The actual labor of such a radical custom is only a fraction of the time spent making templates of cardboard, measuring, fitting and trying.

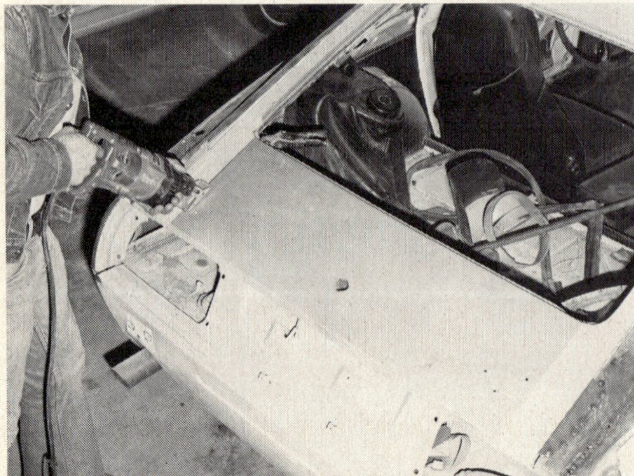

15 The Milwaukee Sawzall is used again here, as attention now is focused on the new rear hatch to be built, which will have a Pantera/Ferrari-type window with sail panels. Rear glass is removed and panel cut almost all the way to the rear.

16 This panel is bent down with just the rear edge intact. The customizer uses a PRO angle-finder/level and a straightedge to determine the right amount of bending to form the rear edge of the new hatch opening, to be metal.

17 Masking tape, cardboard, ruler and measuring tape are used frequently to previsualize the effects of desired sheetmetal changes. Here the customizer is laying out the design of the new hatch. Note the lid is still tacked shut.

18 Steel tubing (½-in.x½-in.) is used here to build a framework for the new sheetmetal and rear window. The window frame is made symmetrical by making it in two halves, hot-bending both at one time over a 2½-in. pipe.

19 The method may seem crude, but it works. The metal skin for the new hatch is bent by hand over a length of 2½-in. pipe. The wise philosophy that governs this and most other kinds of customizing work is "measure twice, cut once."

20 The virtue of patient, careful measuring and bending is rewarded when the new metal fits the opening like a factory piece, with just the desired tapering curve at the sides, so the new sail panels are neither too fat nor vertical.

21 Many clamps are always used to locate a part during welding. Although you can just close your eyes for doing instant spot welds, shield is used for continuous welds. Hatch "skin" has been cut down to mate closely at sides and rear.

22 When welding any long edges such as this hatch, the tacks should be placed about 4 ins. apart until you've gone all around. To cut warpage, hammer as you go. Spot weld every three ins., then every two, etc.

HOW-TO: RADICAL CUSTOM FROM A WRECK

23 The moment of truth comes when the spot-welded patches at the rear are removed and the new hatch opened for the first time. The gap around the hatch/body line has remained perfect. Now all that remains is the window.

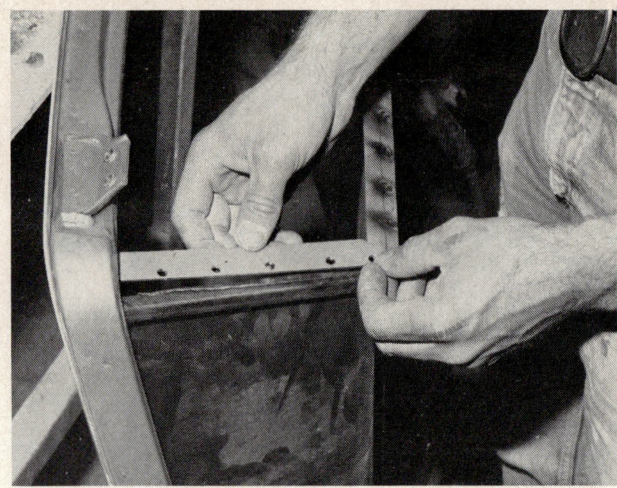

24 The sheetmetal was left long at the front, and was bent over the ½-in. tubing and trimmed. Strips of 16-gauge steel 1 in. wide were cut and drilled to serve as rear window channels. Welding through the holes glues everything together.

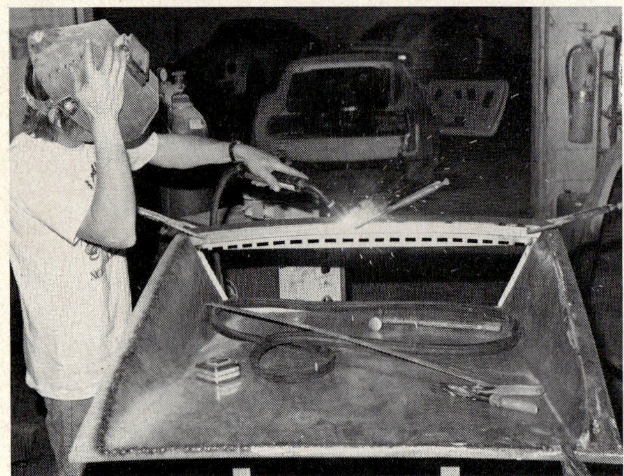

25 A wider strip of 16-gauge was cut to the top curve of the hatch. To make the rubber channel fit where there wasn't a right angle (such as at the top), a length of ½-in.x 1-in. tubing was welded on and top trimmed to a curve.

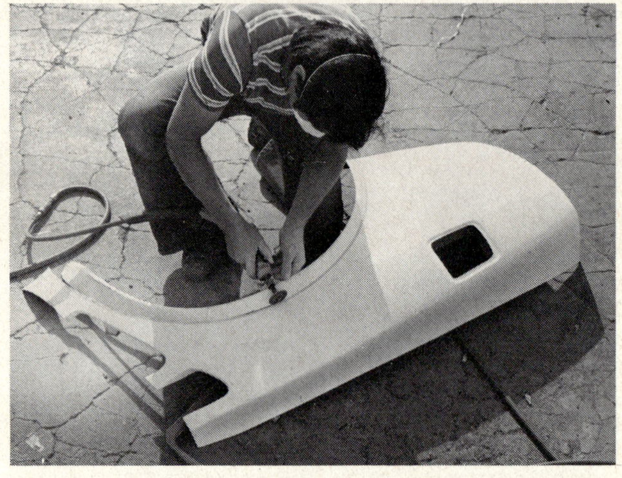

26 We're now ready for installation of the four fenders from Jim Cook Racing. Since this Z already had CGE fender flares in the rear, the new fenders were cut to allow the flare to stick through. A die grinder is used to cut a new opening.

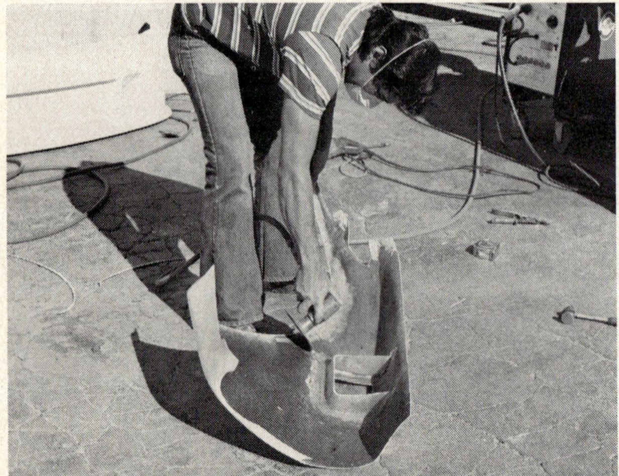

27 The rear fenders were installed with CG Epoxy, which is especially for mounting fiberglass to metal. All the mating edges of the car and the inside of the fenders have to be ground clean and roughed up before using the epoxy.

28 All of the areas on the quarter panel where the new fender will touch are ground and coated with epoxy. In this case, the extra epoxy was spread over all of the finished and ground welds, where it will act to seal any pinholes.

29 The epoxy begins to set up soon after mixing, so several helping hands are necessary to hold the fender exactly in place. One helper then drills the holes, while the customizer follows using rivets to hold the pieces together during curing.

30 Since here we are attaching fiberglass (fender) to fiberglass (flare), fiberglass matte and resin are used where the flares stick through. The CGE flared quarters provide the fullfinished wheelhouse the Jim Cook fenders lack.

31 A very thin wipe of body filler is applied over the reworked quarter panels to cover the filled-in quarter-window and blend the line where the fenders attach. Welds must be sealed with epoxy to avoid cracks.

32 The Jim Cook rear fenders have two scoops at the front, but in this case our customizer eliminated the lower one and made the top one larger, using fiberglass matte and resin. The Western wheels were mounted with PSI adapters.

33 Carl and fiberglass expert Brian Gonzalez (left) contemplate the fit of a Z Products apron from a Group Five Porsche. The fender here has been extended 6 ins. with cardboard and tape. The Porsche apron had to be modified.

34 To stiffen the entire front end and mount the Porsche apron, a steel framework was built from ⅛-in. straps and ½-in. diameter metal conduit, which tied both fiberglass fenders into a unit. The conduit is braced by tubing (arrows).

HOW-TO: RADICAL CUSTOM FROM A WRECK

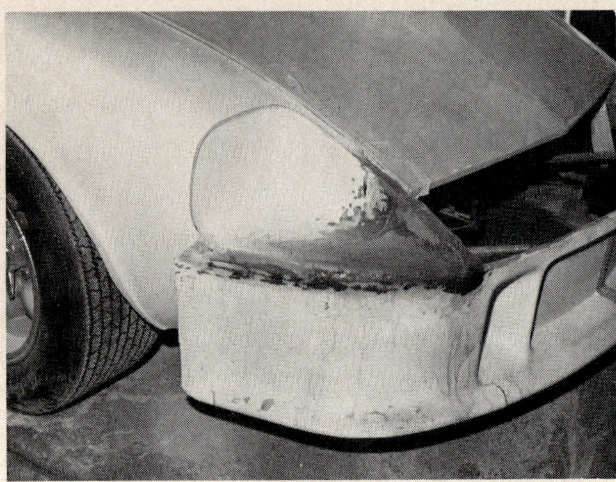

35 After both the fender and the apron are roughed up with the grinder, fiberglass matte, cloth and resin are used to put into reality what the cardboard had indicated. Note gap between the cut-down apron and lipless fender.

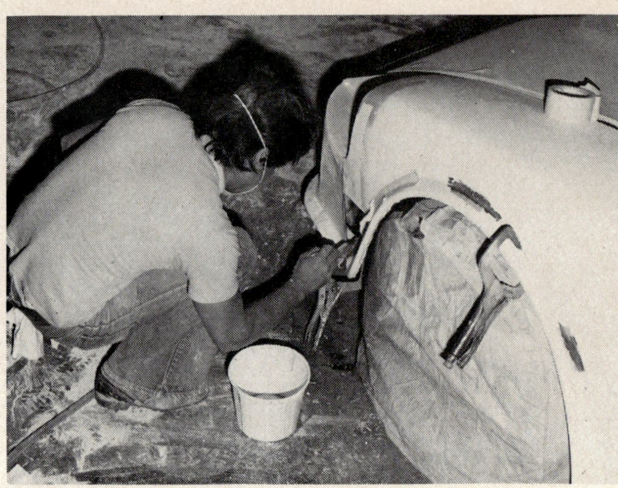

36 Fiberglasser Cuthberto Ruiz dabs in the resin. The front fenders had no lips, so the flared edge of a pair of CGE fiberglass front fenders was grafted to the Cook units. Clamps and small strips of 'glass hold it for checking fit up front.

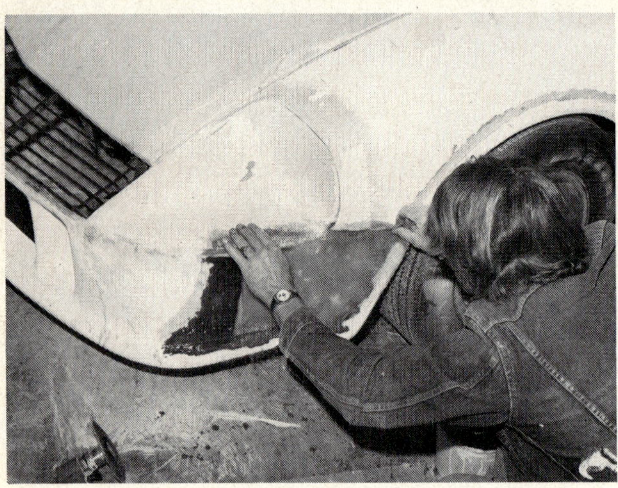

37 With the fender lip completely 'glassed on and the fender extended up front, the lower/rear portion of the Porsche apron was sectioned to keep an even line along the bottom. Fiberglass over cardboard fills in the corners.

38 Barry Kratzer of Arrow Glass has experience in chopping windshields, a delicate job. He cuts pieces of the stock Z windshield rubber and places them around the new opening to act as spacers and to indicate the cutting line.

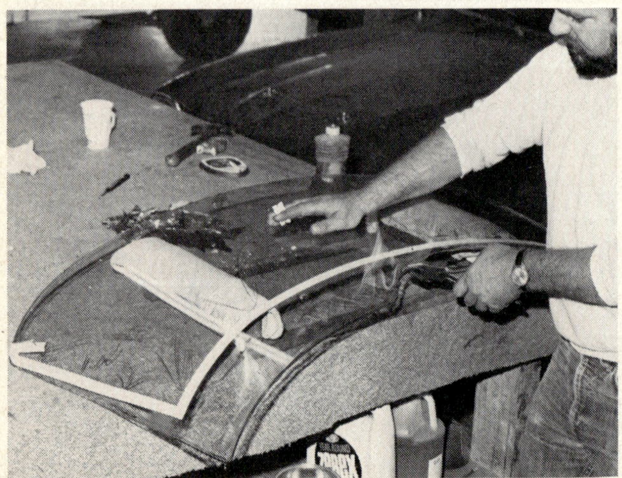

39 In small bites, the glass is scored on both sides with a glass cutter, and broken off with pliers. Burning alcohol warms glass, softening plastic inside for cutting. Final edge of the new shape is arrived at on the water-cooled glass grinder.

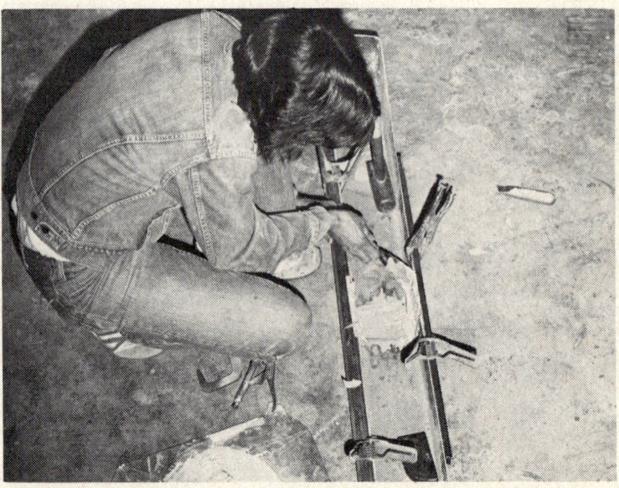

40 Body filler and fiberglass applied at the same time will cure together rock hard. Brian Gonzalez widens a fiberglass Porsche 911 rear bumper in a temporary sheetmetal mold. First filler was applied, then fiberglass immediately after.

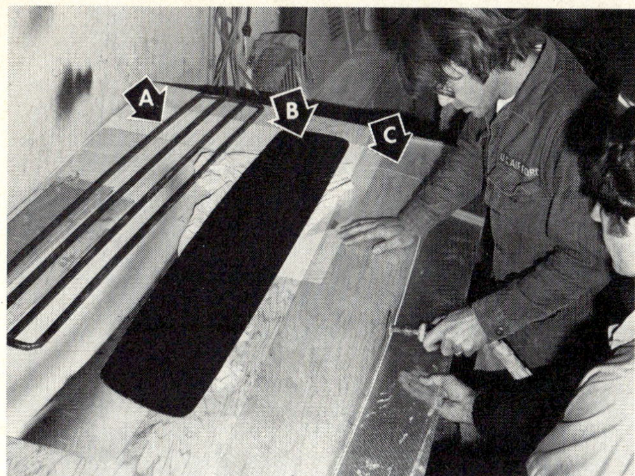

41 New, full-width taillight buckets were fabricated from sheetmetal and installed in the body. The outer lens was made from ½-in.x½-in. steel tubing framework (A), backed by red plexiglass (B) and white Reflex plastic (C) to spread light.

42 The light is backed up by shiny aluminum and eight bulbs. Quite an effect at night! Above it here is the shortened Jim Cook rear wing, with the edges painted flat black to simulate rubber. Porsche 911 guards will be used.

43 The rectangular headlights are 1978 Ford Fairmont, with a black molding made of plastic interior windlace. Fiberglass was used to take a mold from a stock rectangular bucket and these buckets were fiberglassed into fenders.

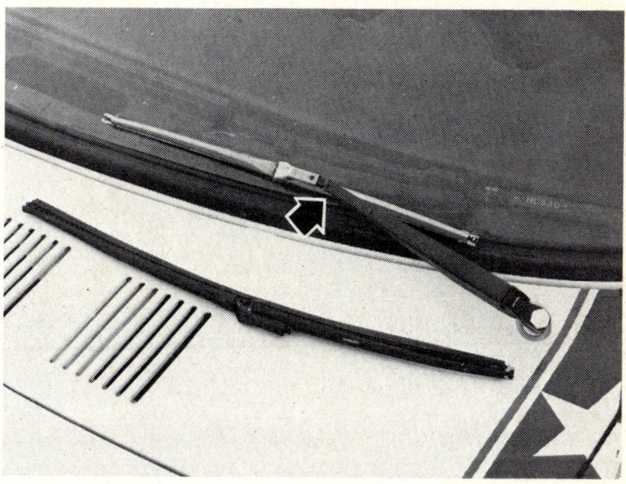

44 The shortened top necessitated some shorter windshield wipers, too. In the case of our Z project, the arms were cut down 2 ins. and rewelded (arrow). The stock Datsun 18-in. blades were then replaced with 12-in. blades.

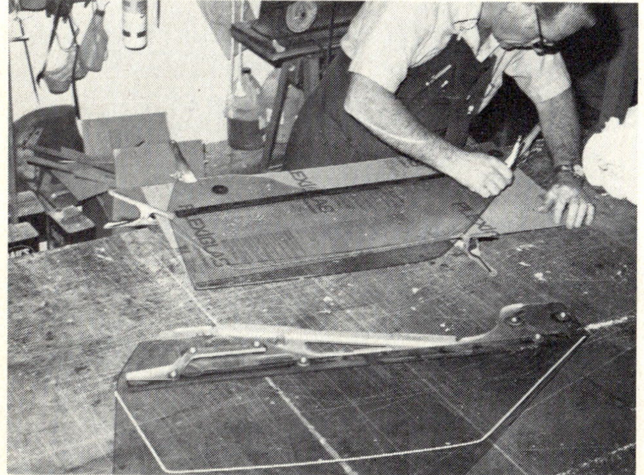

45 George McMillen of Gem-O-Lite Plastics Corp. made new side windows from smoked acrylic, since Herculite side windows cannot be cut down. He followed the ⅛-in. tape line Carl made on the stock glass. Note new angles.

46 White rubber molding from a hardware store makes the front end appear even lower, but flexible rubber is practical for everyday driving. See the completed radical Z with Evel Knievel paint scheme ready for Oakland on page 86.

HOW-TO: VAN CUSTOMIZING IDEAS

The growth in enthusiasm for vans and in van sales has been phenomenal the past few years. Van sales are increasing at a faster pace than any other segment of the automotive market with the exception of four-wheel-drive vehicles.

Part of the reason for van popularity is undoubtedly durability, since a van uses many truck components, and part is utility, for a van combines very well the functions of a station wagon and light truck. But primarily the reason is simply that vans are fun vehicles. This being so, the popularity of vans is not going to diminish in the forseeable future, unless their prices rise very steeply or governmental regulation snarls them in red tape.

By their nature, however, vans look alike as peas in a pod. Human nature being what it is, vanners naturally like to change this, so vans are a fertile field for customizing.

Vans lend themselves extremely well to customizing. The broad flat sides simply beg for creative endeavors in the same manner that an empty canvas beckons an artist. Because a van's shape is so basic and the factory color schemes so prosaic, a small amount of customizing make: a notable difference.

For anyone familiar with customizing techniques on automobiles, working on a van will seem easy. The boxy shape means that there are no subtle compound curves to complicate the reshaping of body panels. There is one major difference, however. Customizing an auto almost invariably means removing something, whereas customizing that plain van usually means adding something.

Whole books are available which cover van customizing, so what we present here is a brief overview. On these two pages are a potpourri of ideas which we trust will stimulate you, while the following six pages are devoted to how-to's explaining four specific modifications.

1 A simple blackout of the grille is the single most distinctive change to a van's appearance. The rectangular headlamps, flares, and spoiler on this Tradesman are almost unnoticed compared to the grille's dramatic statement.

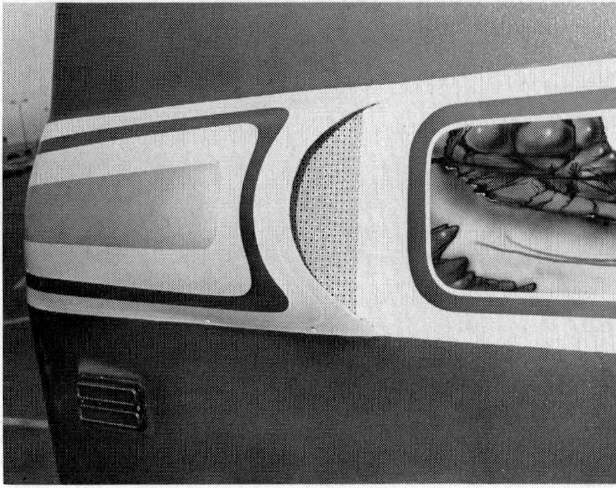

2 An air scoop originally intended for use on an auto hood was adapted for this van. Complete with a section of grille work of the type available at home improvement stores, it channels fresh air into the van's interior.

3 Radical late-model Chevrolet van has a wealth of ideas. Running boards add a touch of nostalgia, and those gullwing doors are knockouts. Though hinged at the top, those doors actually lock and latch at the normal position.

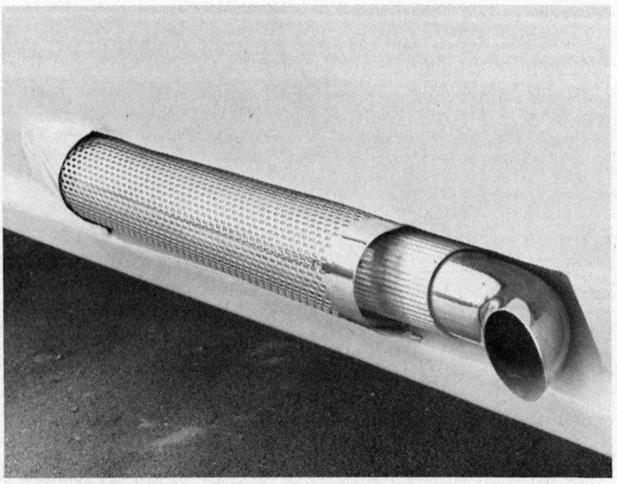

4 Side pipes add a nice touch to any van, but are often awkwardly placed right underneath a door or mounted down low where they reduce ground clearance. This frenching treatment neatly solves both problems with a classy touch.

5 For any customized van with rear fender flares and a sliding cargo door, an extender bar is a must item or the flares will not stay intact for long. Fords use a centered extender, while Chevrolets have an extender at the top.

6 As this early Chevrolet van shows, in some cases it is not necessary to swap taillights in order to obtain that much-desired custom look. Small-diameter steel tubing was welded around the stock light and then smoothed out.

7 Adapted from an unknown source, this twin-nostril hood scoop assures bug-free fresh air for the van's engine. Many imported cars and early muscle cars have hood scoops suitable for vans. Openings can be made to suit the owner.

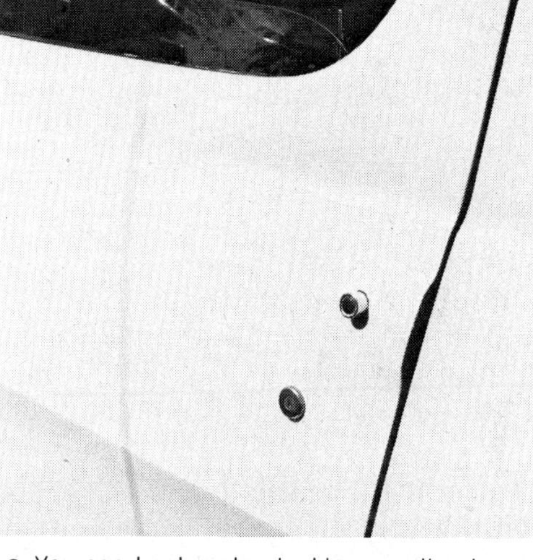

8 You need not go to electric operation to get a smooth surface for the doors, as this Chevrolet illustrates. A plunger button was adapted from another model, while leverage is provided by the finger slot at the door edge.

9 For thinking ahead while looking back, a Mercedes rearview mirror can't be beat. Much more attractive than those huge truck-like affairs that dealers insist on fitting to vans, mirror is adjustable from inside and protects pedestrians.

10 Several accessory manufacturers offer spoiler kits, but this one was fabricated from patterns. The base plates were blind riveted to the roof. The effect is enhanced by the clever continuation of the paint pattern into the spoiler.

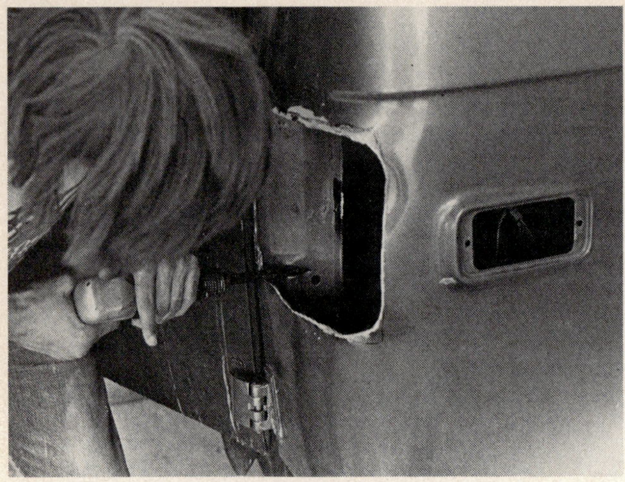

1 Swapping taillights is an effective customizing trick for any van. You can put practically any taillight in any van, but here we show late-model Pontiac Grand Prix lights in a Chevrolet. An air chisel roughs out the hole.

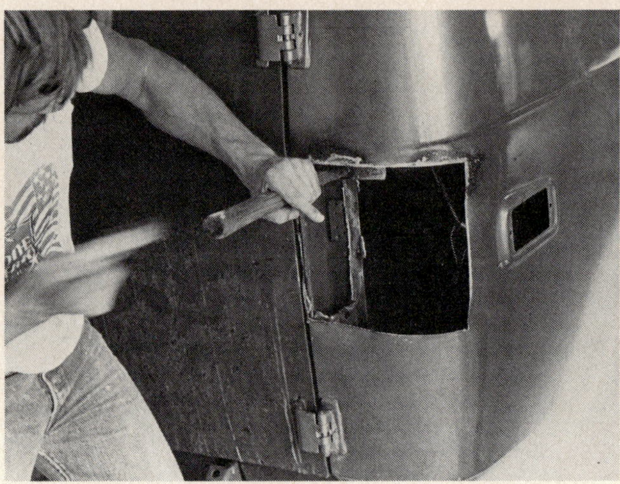

2 After squaring the hole edges with a hacksaw, a torch or air chisel is used to cut away part of the vertical rib just beside the rear door. This is necessary to create rearward clearance for the Grand Prix taillight.

3 Trial fitting is necessary at frequent intervals to insure a perfect fit, even though some fiberglass will be used later. Faceplate should be fitted from the rear and should contact the edges of the hole cutout on all four sides.

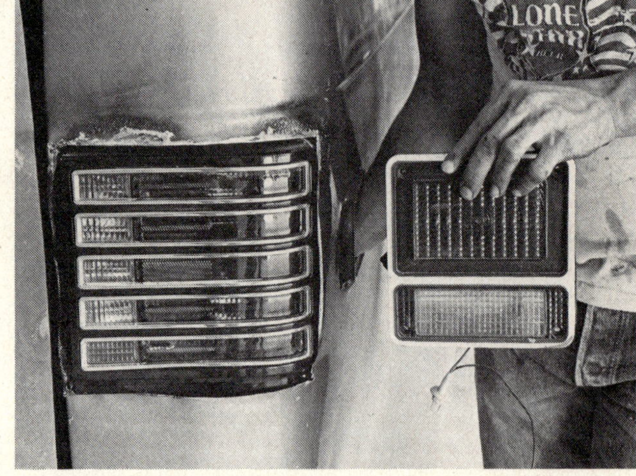

4 Comparison shows difference in size and configuration between stock taillight and Pontiac unit. Difference is very noticeable at night, especially to owners of other Chevrolets. Same bulbs and wiring are used.

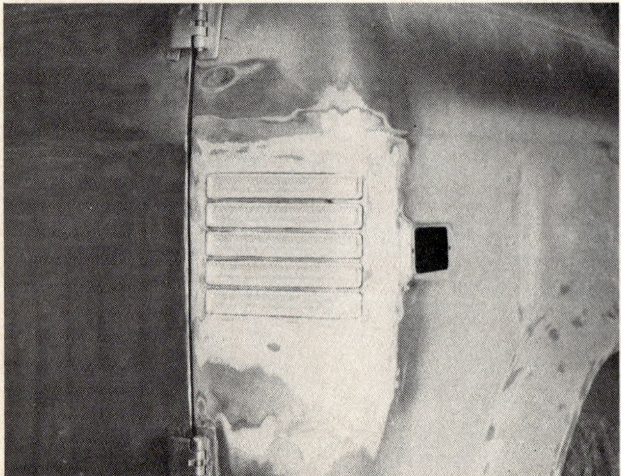

5 Because faceplate of Pontiac light is plastic, fiberglass was used to bond the faceplate to the van's sheetmetal, effectively sealing the gap between them. Masking tape covers the lenses during the glassing and painting operations.

6 A little body putty, a little primer and sanding, followed by a decent matching of finish paint with the original color, and the job is done. Note how well the Pontiac taillight matches the stock side marker light.

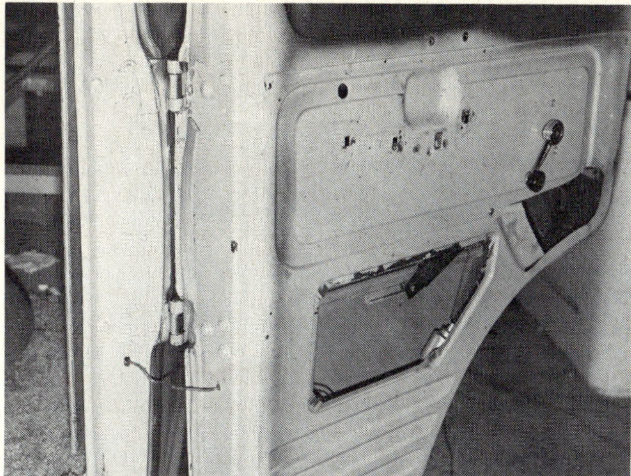

1 Though it can be performed on any van, this conversion was done to a Ford Econoline by MRK Vans. Hinge relocation required adding a ⅛x2-in. steel plate to each door and pillar. Doors had to be notched for hinges.

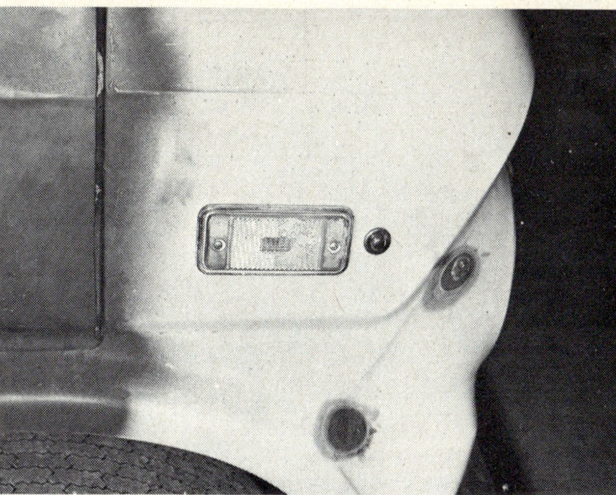

2 Doors are electrically operated in this conversion, which eliminates the need for door handles. Exterior solenoid buttons are near the side marker lights, interior buttons are on dashboard. Stock door locks were retained.

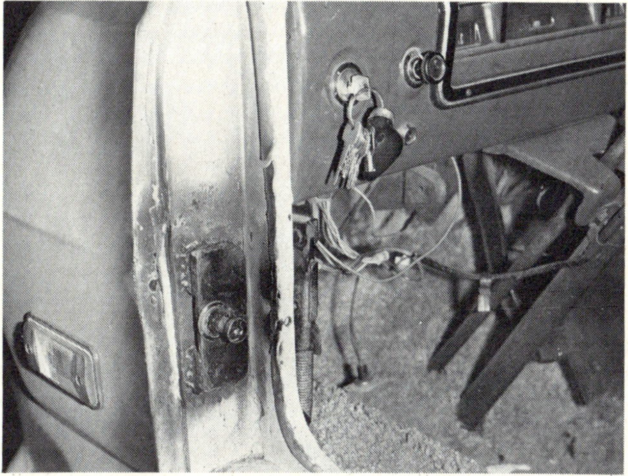

3 The striker plates were aligned to the door with ½x1½x1-in. steel backing plates and then welded. These are stock units; other combinations were tried, but MRK found that the original type of latches and strikers were best.

4 Door latches were relocated from side to side; former left door latch now went to right door, and vice versa. Reinforcing of the door in this area did not prove to be necessary, and latches mounted in stock manner.

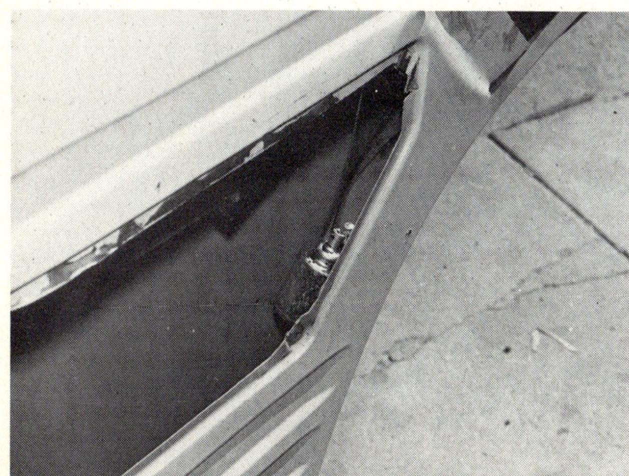

5 A 6x7-in. hole was cut in each door to provide access for the installation of a Cal Custom electric door solenoid. These solenoids were the only non-stock parts used in the conversion, and are not necessary if stock handles are used.

6 All finished except for paint, the converted Ford van shows off its rehinged doors. Exit and entry are now easier. Total cost, including labor, was $500 per door. MRK has plans to market a conversion kit which greatly reduces this cost.

1 A 7-in. chop on this Dodge Tradesman begins by outlining the area to be removed with tape. A rough-cutting air chisel can be used since the skin and ribs will be lap-welded, not butt-welded. Cut the outside skin only, leaving ribs intact.

2 Customizer then marks the crosscut line in the roof, just ahead of the frontmost roof recess. Note that the windshield is already removed. Cut across roof, then cut side ribs to release roof from body.

3 After roof removal, the section over the windshield is then removed by cutting high on the windshield posts. This small section is then put aside for the time being and work is concentrated on the rear roof panel.

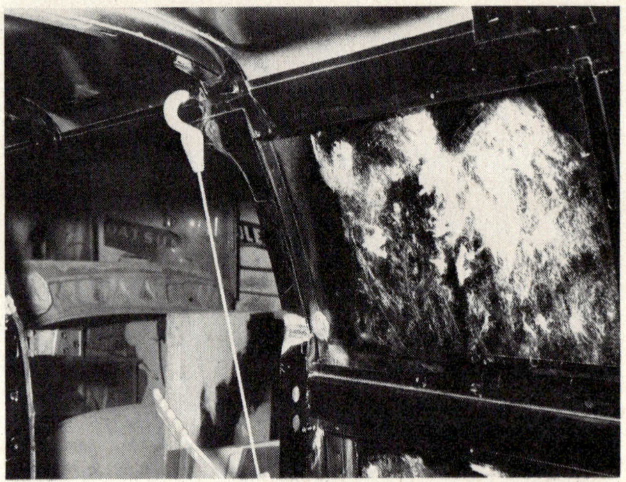

4 The roof is lowered into position, using the ribs as line-up points. Slip the upper ribs inside the lower ribs and the upper skin inside the lower skin. Use a cable come-along to pull the roof down. Alignment at door jambs is important.

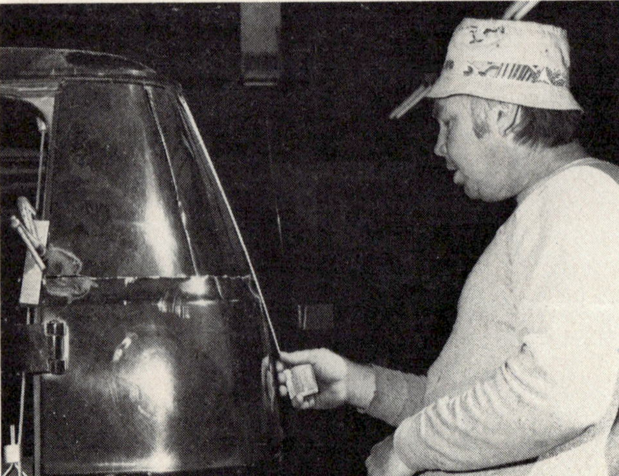

5 To make sure that the roof is level all the way around, measure from roof gutter to lower body character line. On this particular chop, this distance was 14 ins. Allow no more than ⅛-in. variation. Now tack weld the ribs.

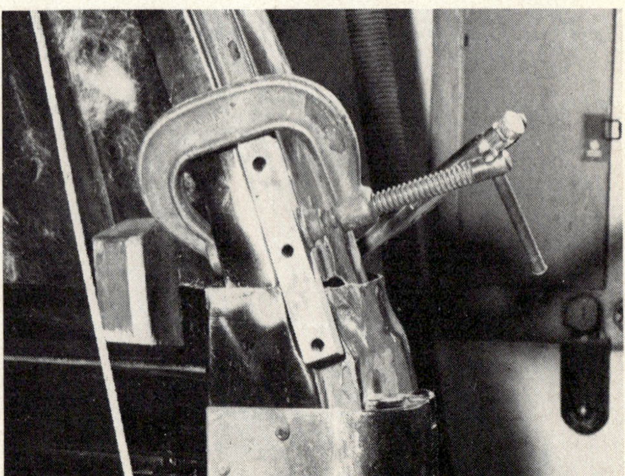

6 Cut each rear quarter of the roof from the bottom to the rain gutter. Pull each rear section of the roof rearward to align it with the lower sections of the rear doorjambs. Clamp these pieces into position and tack weld.

7 Cut two triangle-shaped pieces of metal to fit the gaps left in the rear quarter panels by the previous operation, and tack weld them into position. Weld the ribs and doorjambs solid. Now hammer-weld the rear quarters.

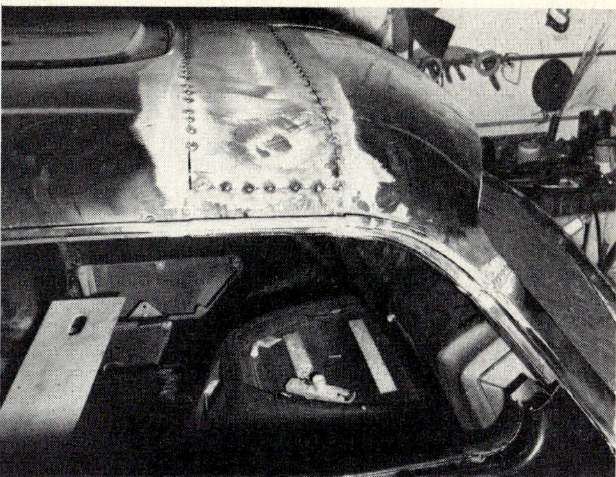

8 Now cut the windshield posts and tilt them inward to mate with the small roof section you laid down a while back. A strip of metal must now be cut to fill in the gap between the roof and forward roof section. Hammer weld in this strip.

9 The unfinished welds here show the cuts necessary on the inner surface of the side door. Some involve lowering the window opening, but most were needed to relocate the window tracks inside. Here's where real professionalism helps.

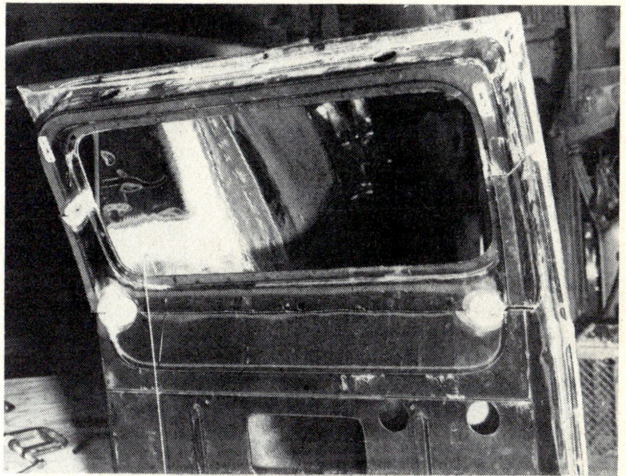

10 Both rear doors were first cut 7 ins. in the window area, then leaned inward to match the new roofline by making a cut just below the window area on the inside of the door. Ditto for the side doors, then they were welded.

11 The stock Dodge seat pedestals were cut down so that adequate headroom would remain under lowered top. The amount of cutdown will vary according to the owner's height. Tabs were then welded on for stock seat tracks.

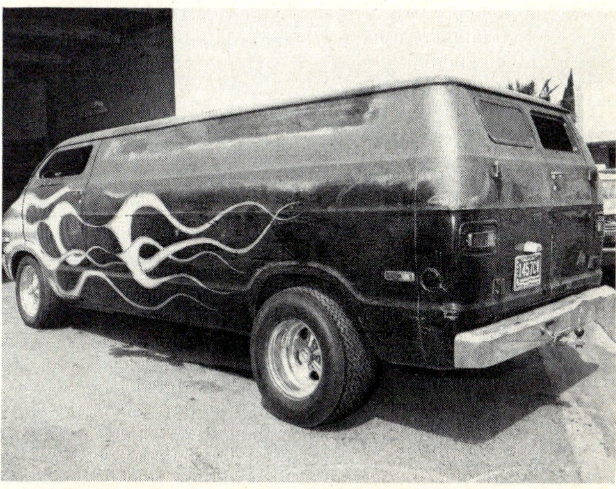

12 Finished except for repaint and installation of cut-down glass, the lowered Tradesman shows how much sleeker a top chop can make a boxy van look. It's an expensive proposition, but nothing else will add so much to a van.

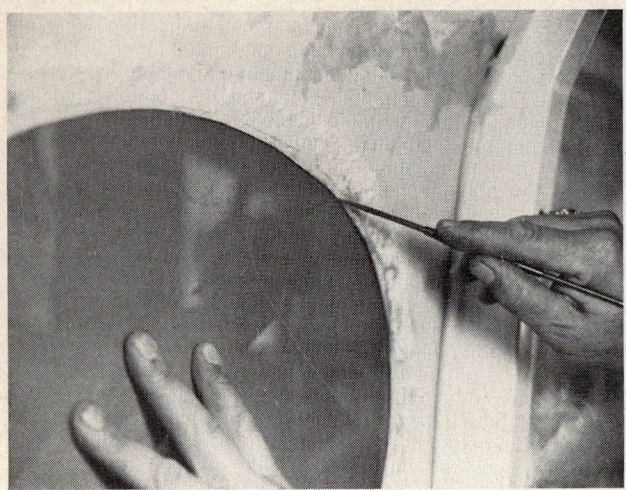

1 Times and tastes change, and the multitude of portholes which sprouted in recent years are losing favor among many vanners. They can be filled in. Remove the porthole; scribe a line from inside on sheetmetal held from outside.

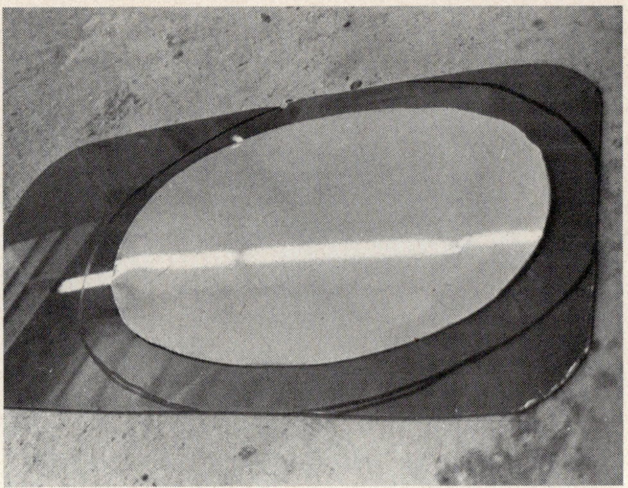

2 The sheetmetal should be the same thickness as the van's body sheetmetal; it's best to use a micrometer to be sure. With tin shears, follow the scribe line to cut out a filling plate, then cut out a backing plate 3 ins. wider all around.

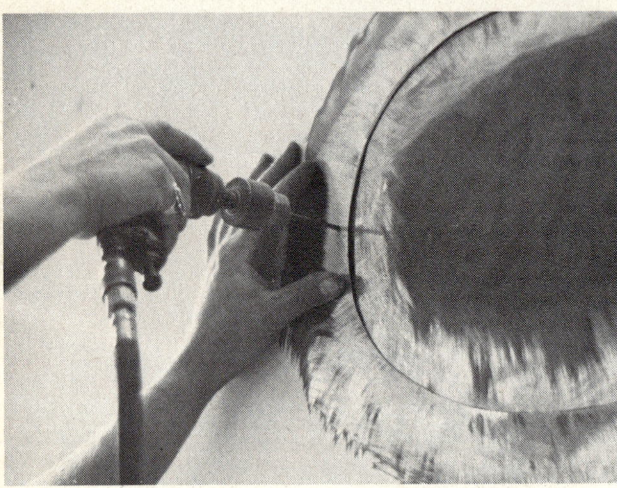

3 With an assistant holding the backing plate from inside the van, drill holes through the van's side panel and the backing plate. Drill only a few widely-spaced holes at first, as fasteners will be used to hold the panel before more drilling.

4 Cleco fasteners, which are removable, work very well to hold the backing plate in place temporarily. If you cannot find these, pop rivets can be used, but they will have to be drilled out later. Sheet metal screws will also work.

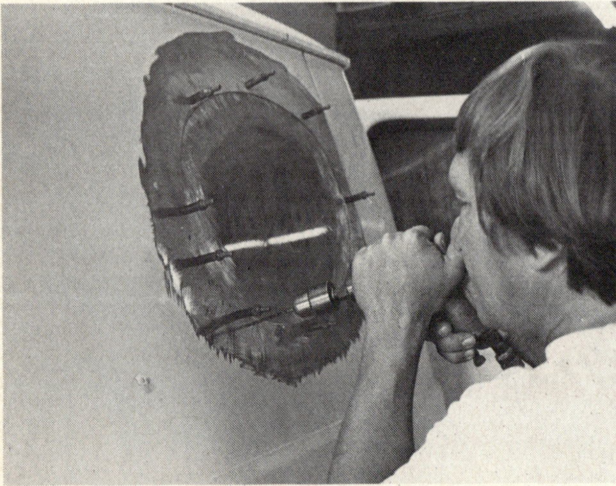

5 With the backing plate held in place by several fasteners, proceed to drill holes completely around the periphery from the outside, spacing the holes at approximate 1-inch intervals. Exactness of spacing is not important.

6 With the backing plate now drilled and held snugly in place, the hole filler plate is inserted in the space once occupied by the porthole. Since it was cut out of metal exactly as thick as the van's sheetmetal, it should fit precisely.

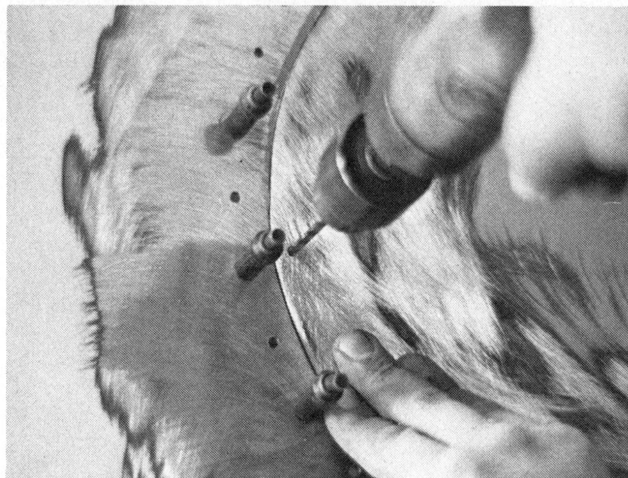

7 Now the hole filler plate and the backing plate are drilled through in the same manner as was done to the van's side sheetmetal. Again, Cleco fasteners, pop rivets or sheetmetal screws can be used to hold the plate for drilling.

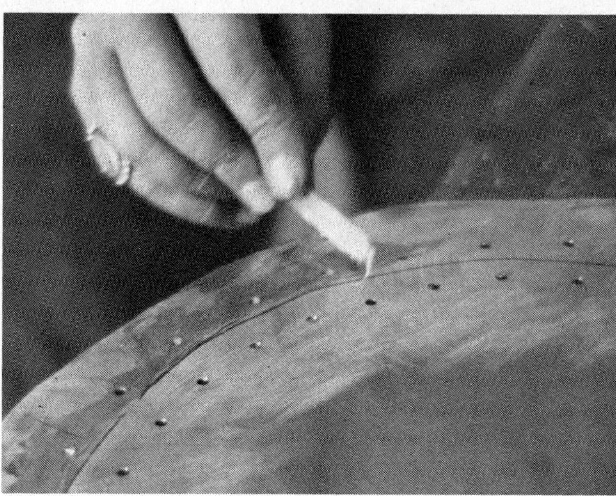

8 Once all holes have been drilled, remove all fasteners and lift the two panels out of the van. De-burr all holes and edges of both panels, then countersink all drill holes in the filling plate and the side of the van.

9 Now it's reassembly time. After running a generous bead of epoxy around the edge of the backing plate, put it back in position and install pop rivets in all its holes. It helps to have an assistant hold the plate until several rivets are in.

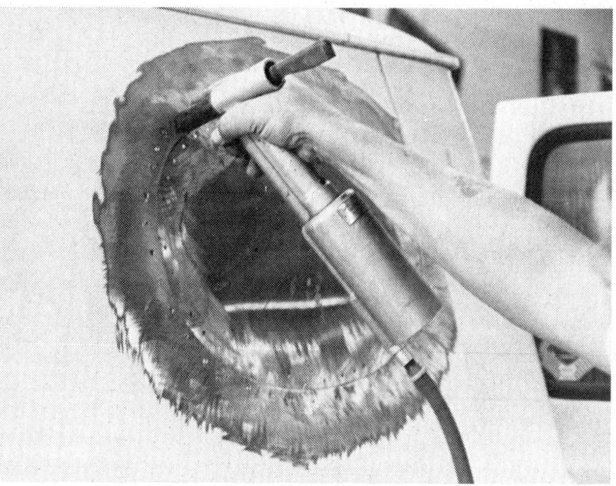

10 Since all the drill holes are carefully countersunk, the tops of the pop rivets will be flush with the side panel metal. This conversion has been a lot of work, but now the results of this elaborate procedure are beginning to be.seen.

11 With the backing plate now secured, apply a smooth layer of epoxy to the back side of the filler plate. Position this plate in the hole and secure with rivets as was done with the backing plate. The filling job is now complete.

12 Working time to this stage is about four hours. The panel is now ready to be finished off by grinding, sanding, putty, primer, and paint. This panel is now strong enough to be cut into if the installation of a vista window is desired.

by Eric Pierce

Though some well-known customizers subscribe to the philosophy of, "If you're going to chop it, chop it good," meaning a healthy whack is preferable to a mild one, it was decided to remove only 2½ ins. from the top of this LUV truck. A conservative chop like this often goes unnoticed, but it is truly appreciated by other LUV owners since it greatly enhances the truck's appearance (rather than making it look grotesque.) A mild chop also retains an adequate amount of headroom for people of average height.

The key to simplicity in our project is a strip of steel 4 ft long and exactly 2½ ins. wide. This is used to measure and scribe the cuts so that they are exactly parallel. All cuts are made parallel to the tops of the door frames and are best measured with a gauge level like that shown in the photographs. First the two doors are cut and rejoined using a piece of the material removed to extend the top section so it fits between the severed sections. The turret is removed next, severing it from the body at the top cut. A 2½-in. section is then sliced from each of the four posts, taking care to make the second cut at exactly the same angle as the first, or top, cut. The turret is now lowered into the position it will later occupy. It will sit on top of the post stubs only on one side and must be propped into a similar attitude on the other because it is now too narrow to fully bridge the gap. A careful measurement is taken from the bottom center to the top center of both the windshield opening and the rear window opening. This crucial measurement must be duplicated when the turret is reinstalled, post to respective stub. This determines the proper curve for the top chop.

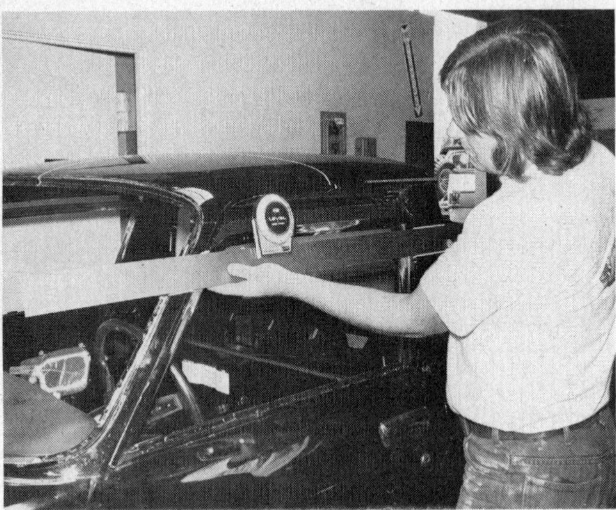

1 Hickey Enterprises' bodyman Joey Allen used a strip of sheetmetal measuring exactly 2½ ins. wide, and longer than the truck's cab width, to accurately mark the horizontal top cuts. All cut lines will then be parallel and equidistant.

2 In deference to other top choppers. Allen cuts down the doors first. Scribe lines were made parallel to door tops, and well below the upper forward curved portion of the door frames, then hacksawed through; removed piece was saved.

3 Naturally, the rear of the door frame was also scribed and cut. Center of door frame was also cut through, for as the frame is lowered to align with bottom stubs it must be stretched, and the section removed from the post was used.

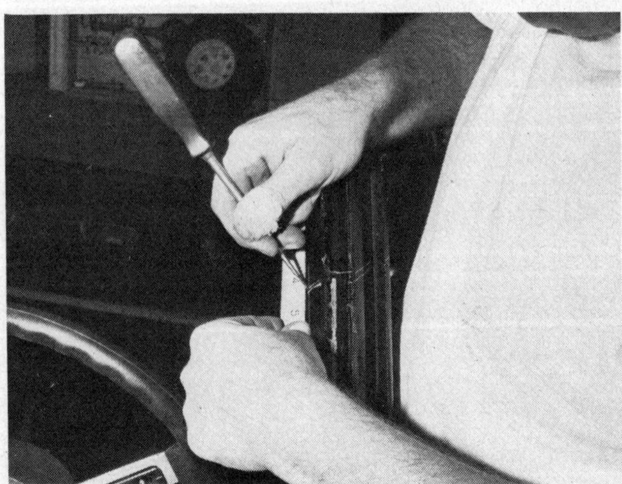

4 Though the sheetmetal guide was used to scribe the cut lines on the windshield posts, it's also necessary to scribe all around the post as a hacksawing guide. If upper and lower cuts are inaccurate, post alignment will be off.

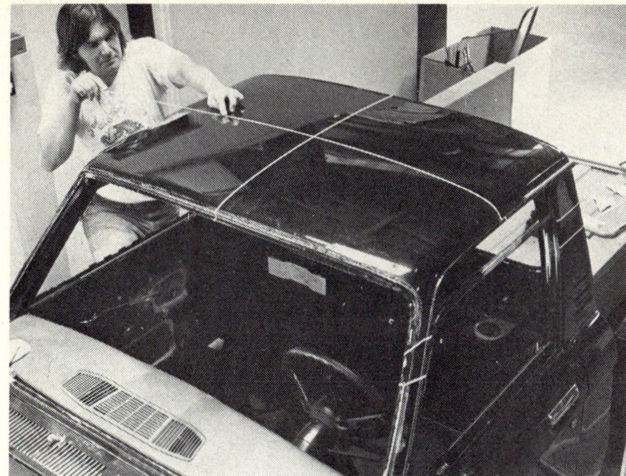

5 Later, the turret top must be quartered (you'll see why as we go along) but it's marked now while the top is still intact and attached to the cab. Cut lines don't have to be precisely located, but they must be straight as a string.

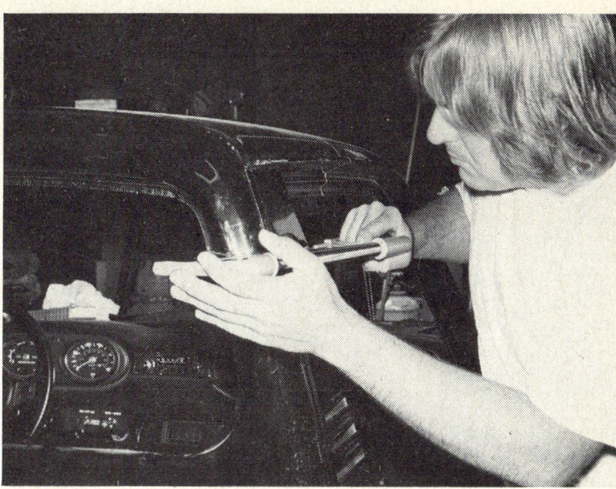

6 Nothing will substitute for the old-fashioned hacksaw in making accurate, clean, and precisely parallel cuts. The four topmost cuts are made first, the turret removed from the cab and set aside, then the lower cuts are made last.

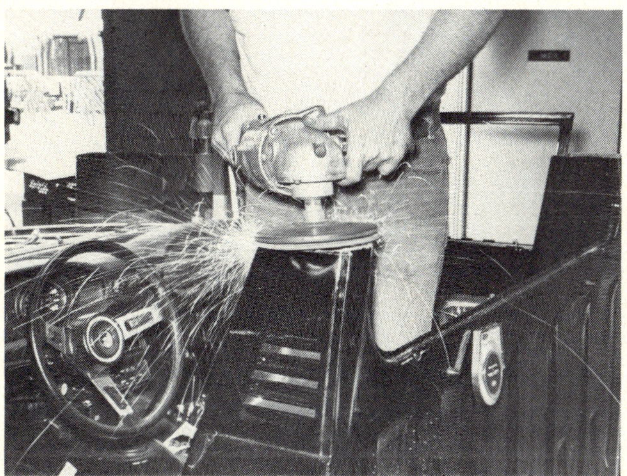

7 A light touch on the stubs with a body grinder will clean up minor imperfections and allow a precise mating of upper and lower pieces, but don't bear down or you'll wind up with a turret top that is lopsided.

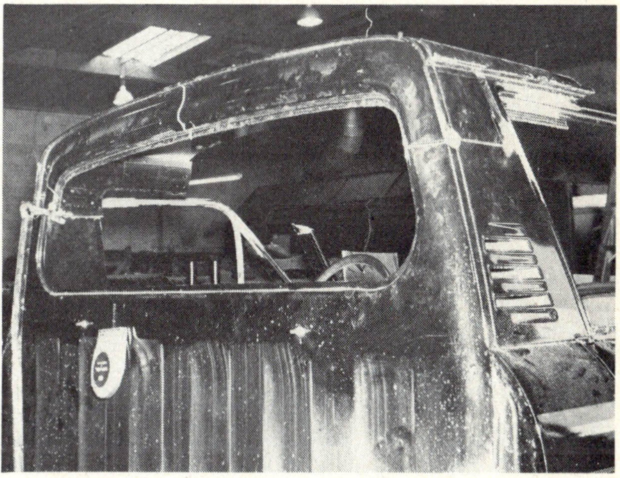

8 Reuniting the top and cab has to begin somewhere, and this corner was as good as any. Spotwelds will hold this section together for awhile. Other posts naturally misalign, but top was propped up temporarily. Now, slice roof lengthwise.

9 Move opposite side of top over until the rear post aligns there, rig clamps to hold center of top in position and across all the gaps. Slice top across the center now; move each forward quarter to posts, align and weld them.

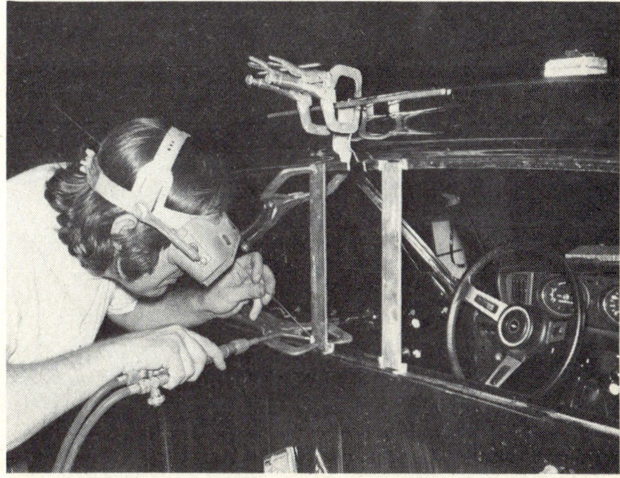

10 Temporary metal supports are spot-welded in window opening for precision alignment. Next step is to bend up a strip of sheetmetal to duplicate shape of upper opening flange, then weld it solidly in position.

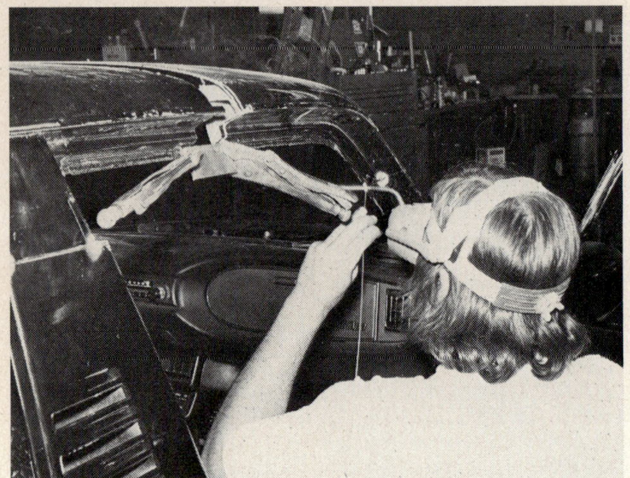

11 A bridge must also be made for each gap above middle of both doors. Shape a metal strip to duplicate as exactly as possible the form and contour of the original doorjamb. Check as you proceed so pieces stay in alignment.

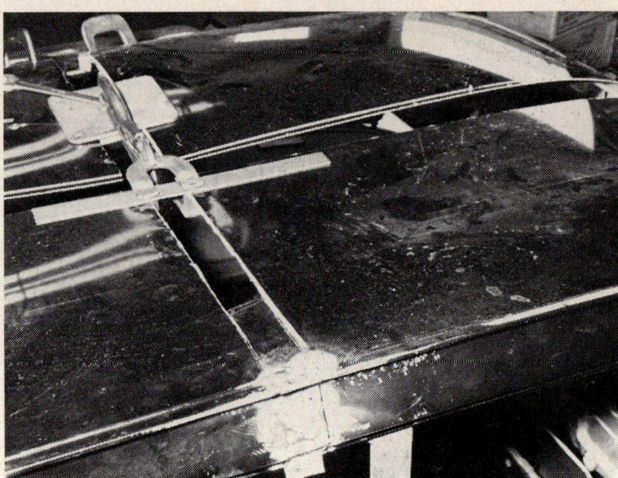

12 Here's another look at the bridge at center of rear window before final welding; extra length at bottom will be hammered in and around to form necessary double thickness of flange. Still-remaining gaps are filled later.

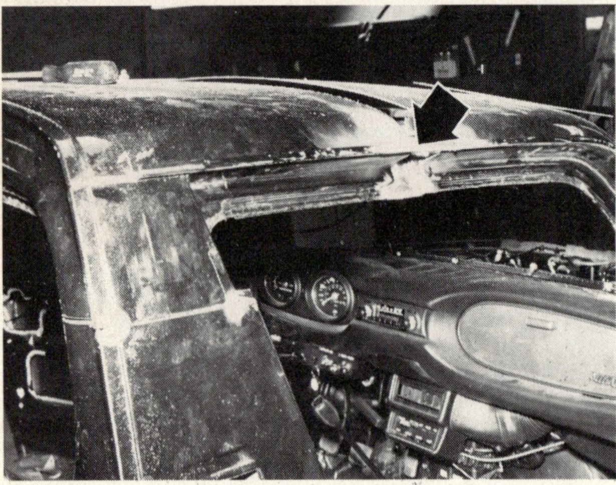

13 Short length of rain gutter is taken from discarded piece of windshield post and added as indicated by arrow. This piece would otherwise be very hard to form to match the original, and would surely catch the discerning eye.

14 Ah-hah; fooled you, didn't we? Instead of having to weld up all the gap lengths that resulted from quartering and spreading the turret top, we're dropping in a sunroof. This will help align the sections and eliminate work.

15 Because the vertical rear panel is tapered, top part no longer matches wider lower part. A piece of sheetmetal is formed and added as shown to make a perfect door opening edge, and the process repeated on other side.

16 Sheetmetal filler strips are made to fill the top gaps, and are at first only tack welded as illustrated. Do the four as shown before solidly welding any of them—don't go ahead and completely finish one side; do them together.

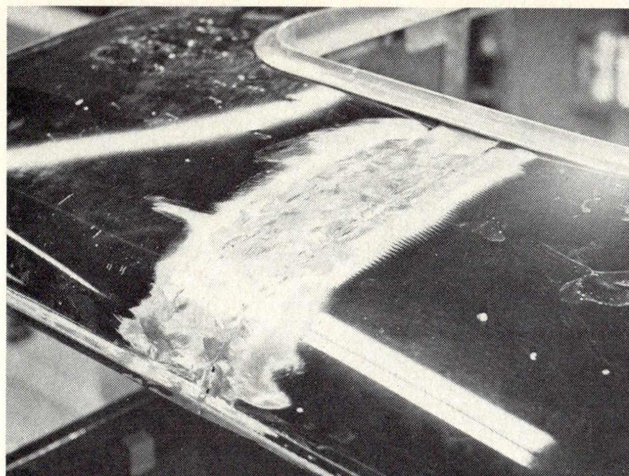

17 Now go ahead and weld up the gaps between each tack, but hammer-and-dolly the seam about an inch at a time while it is still red hot. Do this at alternate spots so there isn't a concentration of heat which causes warpage.

18 When all the gaps are welded and the seams ground down, Bondo or similar filler can now be applied where necessary. If the metalwork is as good as it should be, only a thin skinning with filler will be necessary for a smooth finish.

19 When first coat of filler has set up, file it down; then use more filler if there are any low spots, filing again. Prep the various areas and spray on coats of primer-surfacer, block sanding between coats to work out minute low places.

20 The glass man has done his deed now, and the windshield and rear glass are in place. The doors have been fitted, too, and their glass similarly cut down. From here on out there's more sanding to do and, finally, finish painting.

21 The mini-truck owner was fortunate in one way, he wanted the LUV to remain in its original black finish, so the areas affected by the chop job had only to be spotted in. A light dose of pinstriping really helps to bring it alive.

22 Author Pierce shows off effect of chopping. As noted at the outset, chop was minimal and in keeping with practicality. A greater height reduction would have been just as much work, but result would not have retained proportion.

Fender flaring is now virtually routine work at the busier pro custom shops, projects of the type growing in popularity as states enact laws requiring that tires and wheels be encircled by bodywork. With super-wide tires and deep dish aftermarket wheels finding ever-growing sales, a variety of pre-shaped flare kits are finding their way to dealer's shelves that fit the more popular cars, vans and off-road vehicles. Flares are of either fiberglass or metal, and they can be screwed or pop-riveted in place then left as is or molded to the parent fender material for the one-piece look. On occasion, however, a car/tire combination will show up that requires some dexterous hand-work, as the Sunbeam Tiger illustrated here fitted with B.F. Goodrich T/A Radials. Nothing of a ready-made nature was available off the shelf for this combination, yet the owner insisted on 4-in. wide, all-metal flares on his all-metal sportster. His wide, BR50x13 rubber meant the fenders needed some serious attention.

The proper approach in a situation like this, while rather straightforward for the pro who has performed similar tasks, does require some forethought on the part of the first-timer for the technique is a little tricky at best. The real test is to not allow much fender deformation, requiring extra metalwork which can be costly to the customizer through lost time.

Since fenders have a rolled edge around the wheel opening, which acts as a stiffener and prevents vibration cracks, a similar edge must be formed on the flares. This is most easily accomplished by using ½-in. diameter electrical conduit, known in the trade as EMT, which can be hand-bent to any desirable radius. As shown here, it provides both the outline of the new wheel opening and provides a "sheleton" for the added sheetmetal.

1 A common problem; wide tires, offset wheels combine to put rubber outside of bodywork. Most states frown on this situation.

2 No existing flare kit suited this Sunbeam Tiger, so custom metal flares are started by bending EMT conduit around a tire for right radius.

3 Although pliable conduit will bend to proper radius, it tends to spring outward so crossbraces are temporarily added to hold it at right arc.

4 Conduit is spaced at desired distance from fender and at a point marking ultimate lip of the flare. Line will denote start of flare-out.

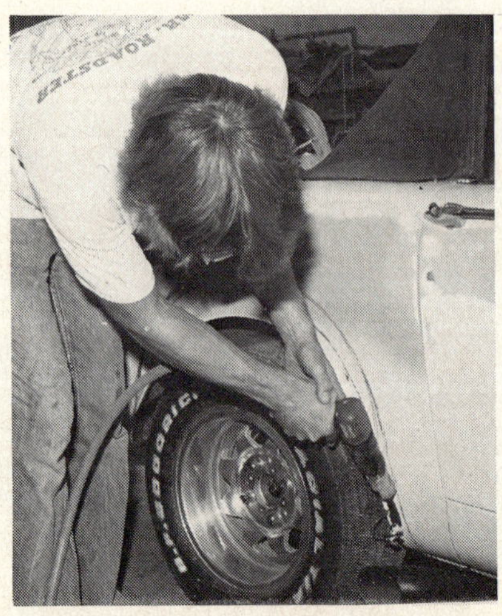

5 Stock wheel well on Sunbeam was double-walled, so air chisel is used to remove original fender lip and inner panel juncture.

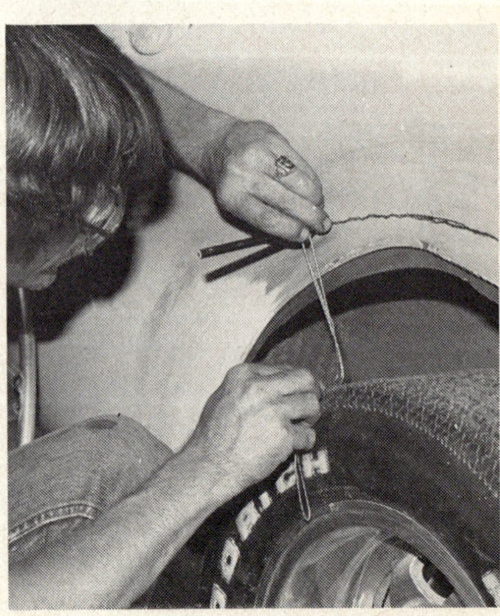

6 A "compass" of sorts is made to trace height required for tire rebound at top of wheel opening. Flare will start at this line on the fender.

7 Tire rebound space is needed only at top of arch, so offending metal is trimmed away accordingly. Inner panel is to be trimmed off later.

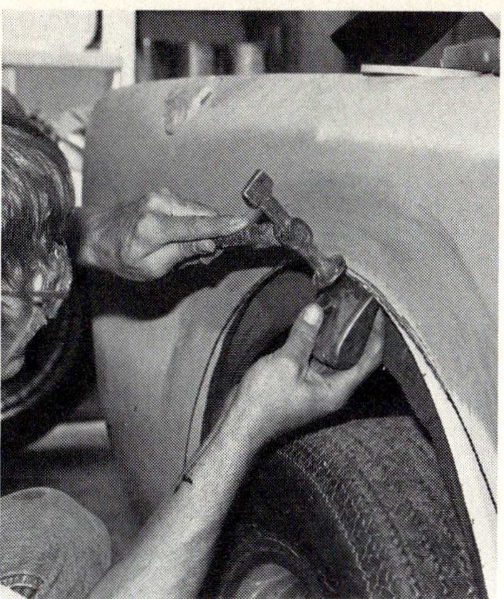

8 Outer fender skin is dollied out to form filleted curve between new flare and original metal. Care here will alleviate severe panel warpage.

9 Conduit "skeleton" is heated to allow sharp bend to bring it in to rocker panel area, where it is brazed at its final desired location.

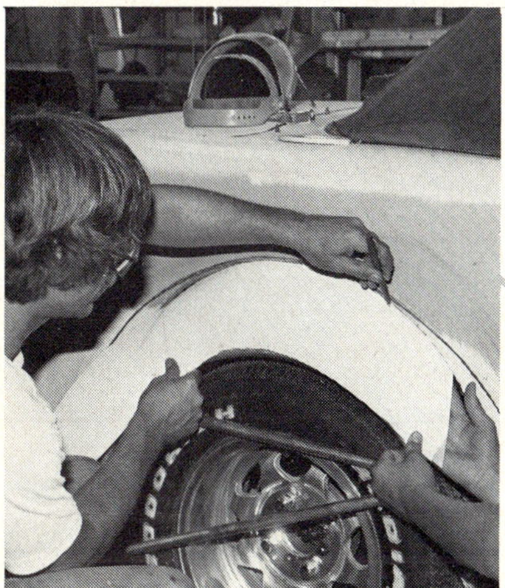

10 Cardboard templates are cut to form bridge between fender and conduit outer lip. Several pieces are made, rather than a large single one.

11 Patterns are transferred to 18 gauge cold roll steel, then pieces are tack-welded to fender, but brazed to conduit framework.

12 Taking shape, rear fender flare still lacks one forward piece. EMT conduit is zinc-coated, but need not be cleaned to bare metal before brazing.

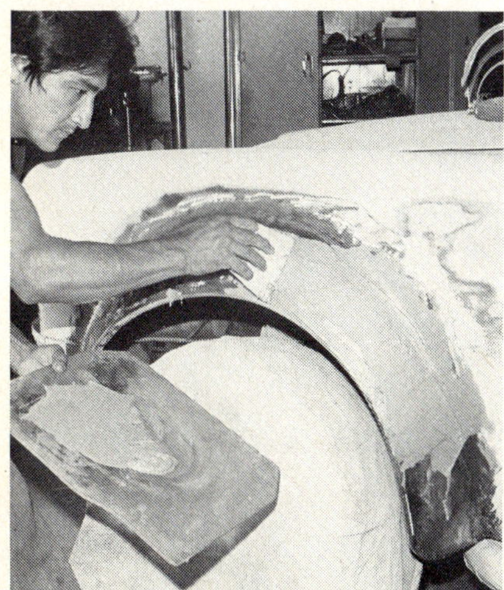

13 Judicious metal fitting, evident in prior photo, left little hammer/dolly work before first thin coat of Bondo is wiped over the flare.

14 Cheese grater cuts away high filler material spots, shows up the lows. A second wipe smooths it out, then usual priming and sanding follow.

15 End result is all the customer could desire; adequate tire coverage and plenty of wheel rebound space. Job ran $900 for four fenders.

Affluency

Conveyances for the socially elite

ILLUSTRATIONS COURTESY PPC LIBRARY

Since the dawn of man's urge to travel he has used his personal conveyance to proclaim his station in life. The rich and the royal bejeweled the chariot, silk-draped the sedan chair, lavishly decorated the rickshaw, gilded the 2-wheel cart, embellished the sleigh, garishly trimmed the wagon, studded the leathered carriage, and festooned the bicycle, all to set their means of locomotion apart from the look-alikes of the commoners. ◻As the engine began to replace the beast of burden, caterers to the carriage trade supplied motorcar bodies to a fledgling automobile industry, at first as the producers of open box-like structures of wood and leather. But the affluent and the elite soon recognized the motorized vehicle as a new object upon which to lavish self-indulgence. Car-body builders, those apart from those who produced thousands of identical units, began handcrafting one-offs to suit individual tastes. ◻Prior to the Depression custom coachbuilders of prominence could be counted by the dozens, but their numbers were severly reduced by The Crash. Yet, a few survived, plying their unique trade until WW II halted flambuoyant mobility. When hostilities ended, automotive uniqueness took a new turn, one where vehicular oneupsmanship could be afforded by the masses. How this evolved is demonstrated here.

The majority of Twenties and Thirities coachbuilders crafted complete bodies to be installed on bare luxury cars' chassis. This Gentleman's Sport Design laid down by Derham in the Fall of 1930 would utilize the 1931 Lincoln undercarriage, presaging the post-war 4-door hardtop convertible. The top was fixed, yet window frames lowered with the glass and the central divider pillar folded down. The final car's total price would have topped $12,000 at the time.

Few coachbuilders chose an everyday, low-cost Ford as the basis upon which to lavish expensive handcrafting, but Jensen Motors of Birmingham, England, pulled it off rather successfully as evidenced by the 1936 Jensen-Ford. It was an unexpected harbinger of what, after World War II, would come to be termed customizing, since it utilized a great many proprietary Ford body panels subtely altered with a sporty flare. Three duplicates of this model were built.

Transition

Crossing the long bridge of time

PHOTOS COURTESY PPC LIBRARY

The long bridge between special coach-built bodies for the great luxury marques of the Twenties, to the full-flowering of low-cost customizing of the Fifties, has seldom been so graphically portrayed as by this Motor Trend Magazine cover photograph for January, 1953. ■ The lovingly handcrafted, aluminum-bodied classic is a 1927 Stutz Bearcat, a 4-passenger boattailed speedster obviously intended for the regal enjoyment of an affluent motorist of the era. But even as costly as such specialty cars were at the time, the price of creative hand labor had risen to such extremes by war's end that such automotive extravagance was virtually dead. In its place rose the art of customizing, where a unique automobile could become the possession of anyone who wanted one by the inexpensive expedient of changing a few design lines of an ordinary car or shifting small parts and trim from one make of car to another. Such is exemplified by this chopped-top 1950 Chevrolet convertible with a 1949 Mercury grille shell and 1952 Plymouth grille bars. ■ The years between these extremes are studied in the opening chapter, and a chronological presentation of the transition of customizing trends up to the present time is contained on the pages that follow.

Two classes of motorists are represented in this scene from 1953 and which leapfrogs 26 years. The aluminum-bodied Stutz Bearcat was a restoration by Elliot Wiener and was intended for ostentatious motoring. So was the Chevrolet of another era, which, with a few body changes and parts-swapping, helped usher in low-cost oneupmanship.

King Kustom

George and Sam "Spell it with a K" Barris

PHOTOS COURTESY PPC LIBRARY

Little can be said about customizing—past, present or future—without inclusion of the name Barris; the good brothers Sam and George, undisputed Kings of the Kustomizers. Ever the showman, often the center of controversy, but secretly admired by his contemporaries, George fronted Barris Kustoms with dash and aplomb. The shop was always staffed with high-ranking craftsmen and were guided by George's brother, the late and great Sam Barris. At the Lynwood, Calif. establishment, it was George who was the front man, making contacts, talking out styling treatments with the clientele, handling the finances and, most importantly, making certain that all work produced gained notariety through car show wins, and that it reached the eyes and ears of the motoring media. He was (and is) a metalworking genius of the highest degree, though at Barris Kustom's peak years during the Fifties, his schedule usually permitted only creation of a styling theme with chalk marks on the shop's floor, or he might rough-cut and bend a car's metal into the rough shapes he wanted, then Sam would take over and see to the final detailing. Barris cars were always unmistakeable, be they Fords, Mercurys, Studebakers, Chevrolets; no matter the year or body style they all bore the Barris *Please turn to page 88*

The most exotic creation of Barris Kustoms during the early Fifties was Golden Sahara. Originally a '52 Lincoln, Barris rebuilt it from a total wreck. Not really typical of shop's work at the time, but later showpieces even outdid this one.

Many were the '49-era Mercurys that Barris Kustoms chopped in the model's heyday, but a rarity among the convertibles and coupes was this 4-door with a typical "floating" grille.

George Barris takes a moment (left) in 1956 to pose with a mildly customized '55 Chevrolet, with stretched fenders flanking a continental kit, while it's parked by a frenched-in Mercury.

Valley's Customs

Pinnacles of perfection from a talented team

PHOTOS COURTESY PPC LIBRARY

No proper discourse on the intriguing evolution of customizing would be complete without mention of Valley Custom Shop, a metal-bending emporium once located in Burbank, Calif., and without peer throughout the Fifties. Owned and operated by the extremely talented brother-in-law team of Neil Emory and Clayton Jensen, it catered to a rather limited clientele of car enthusiasts who knew what they wanted in automotive oneupmanship, but could attain it only through a professional shop's handiwork. Although one could merely have a hood filled or a decklid shaved if he persevered, Neil and Clay much preferred customers who had a basic idea of the overall styling they wished to achieve, but who would let the partners create it without interference and in their own unique way. ■ Valley Custom never equalled the output volume of other famous shops of the time, but most of their creations were so tasteful that they became immediate classics and received widespread notariety through the enthusiasts' publications. Valley Custom's fame was so wide-spread, that a mid-west fan even freighted his '40 Ford's doors to the shop to have the trim and handle holes filled! ■ Though many cars passed through the 2-car garage doors of Valley Custom during the Fifties, *Please turn to page 88*

Valley Custom labored on the Polynesian for seven months, but it was a part-time project and worked on as other shop work permitted. The sectioning dimension was 4 ins., and it could not be taken on a horizontal line as on the '49-era Fords. The '50 Oldsmobile 88 had zig-zag cuts through doors and quarter panels for perfect mating of top and bottom portions.

Ron Dunn's '49 Ford went through Valley Custom's doors twice. Sectioning amounted to 4 ins., as on the Polynesian. When first completed it utilized a close-to-stock grille, but later Neil and Clay reworked it as shown here, further refined the interesting taillight treatment. Just imagine risking superb custom work, unprotected except for skimpy bumperettes.

Dream Truck

A rare pooling of customizing talent.

PHOTOS COURTESY PPC LIBRARY

Thriving mightily to call public attention to themselves, professional customizers of the mid-Fifties pulled every trick of the trade—and invented new ones—to lure customer's dollars to their shops. Through radically restyled vehicles built to custom order, as well as with self-sponsored cars, each tried to outdo the others to win car-show trophies. To do so meant to gain notoriety through the pages of the growing number of car-buff magazines which, in turn, lead to exposure to hundreds of thousands of fans. ■ The better-known customizers of the period were not only hotly competitive but were loudly scornful of each other—professional jealousy personified. Each builder had his own styles of automotive re-design and personal traits, from the flambuoyancy and verve of the great George Barris to the low-key attitudes of Valley Custom's partners who were rarely seen away from their shop. Never, at the time, would one of customizing's leaders work on a car touched by another. ■ It took Spence Murray, however, to prevail upon the hobby's greats to pool their considerable talents for the good of customizing in general. They were asked to jointly create for Rod & Custom Magazine a 1950 Chevrolet ½-ton pickup truck, perhaps the least-inspiring, most

Please turn to page 88

The Dream Truck—once the majority of its bodywork was carried out and the chassis was up on all fours—continued to undergo alteration. In early 1957 Barris had painted it light purple with darker gradations of purple panelling (right). Later that year it was sent to Bob Metz in Indiana where the fins were built and a pearl white paint job added (above). In 1958, just before it crashed, Barris had repainted it in lime gold (top).

Approaching the ultimate in handcrafting

PHOTOS BY ERIC RICKMAN

An extreme in creativity is Proto 1, a superb fiberglass one-off commonly taken for a much-customized Corvette, but the body is Vette only in its rakishly laid-back '63 roadster windshield. Much more than a showpiece, Proto 1 both handles and goes—both attributes of the car's original intentions. Traco Engineering blueprinted and modified the 350 Chevrolet engine, which now produces 450 hp, and a TurboJet 400 transmission drives the machine's all-up weight of only 2298 lbs. The 88.5-in. wheelbase sportster has run the quarter at 116 mph in the mid-10's. ■ A stance of only 47.5 ins. overall height gives less than 5 ins. of ground clearance under the crossmember, while the front air dam just barely clears a cigarette package laid flat. The car has to have a stiff-riding suspension to make it road-worthy, this department taken care of by Vette rear suspension while front suspension, like the steering and the firewall, are Jaguar. ■ Evolution of the striking body during its 4 years of metamorphose is a little fuzzy due to ownership and plan changes along the way. The original concept was, in the vernacular, "eyeball-engineered" by Al Johnson, now of Bruno's Corvettes, who in 1973 began shifting around the basic shape of an old Valkyrie sports car body kit. *Please turn to page 88*

Tony Chirico shows off the low stance of Proto 1; just 47.5 ins. overall height. Design started by the reworking of a Valkyrie sports car body, mid-engined, but was reshaped to form the molds for the all-glass body panels.

The use of proprietary parts is common on hand-built cars, for few builders would attempt to design and build door handles and latch assemblies. Rear glass and panel are Porsche, gas cap is Chrysler, Pontiac door handles.

Long hood was necessitated by engine set back of 35% of the wheelbase. Car's handling is superb thanks to careful chassis engineering by Gary Wilson. All-up weight of Proto 1 is only 2298 lbs.; extremely light. Body was wind-tunneled.

The Evel Eagle

Our Junkland to Oakland project car

by Jay Storer

Daredevil motorcycle riders and automobile customizers are funny breeds to some people because what they love seems "unnecessary" to the narrow-minded. To those people we say "bunk," because we love those fringers for the thrills and variety they put into ordinary living. ■ Seen in this light, the joint venture of premier daredevil Evel Knievel and Carl Green, creator of our customized Datsun project was almost inevitable. Carl has always done the bodywork for Knievel's Stutz convertible and Cadillac pickups, and when Evel saw the super-Z, he wanted it for some of his promotions. The Eagle was born, and with the addition of a Nordskog Chevy V-8 conversion, will soon fly, too! ■ A few photos and some copy can't possibly impart to you the overall impact of this low-down import. Some observers liken it to a cutdown and customized Mustang, others want to know why it's on the street instead of a race car transporter. Little is left of the original Z except the doors and hood; everything else has been creatively modified, added on, deleted (like 4½ ins. out of the roof!), or repainted for the nothing-else-like-it total look. The various components CGE assembled and modified to fit the Eagle attained present unity of design through attention to *Please turn to page 89*

Millionaire stuntmen/businessmen are hard to pin down, so don't look for Evel in the photo at top. He couldn't make it for our photo session the day we shot the Evel Eagle. Daytime lighting hardly does justice to that eight-bulb taillight in the photo above, sandwiched between Porsche bumper and "whale-tail" wing. At right you can get a better look at that hand-painted eagle. design by Bob Gloege. The Can-Am White paint and stars were handled by builder Carl Green.

Wedged Van

New metalworking wrinkles—smoothly.

PHOTOS BY ERIC RICKMAN

Somehow it seems fitting to conclude this particular color section with an outstanding example of today's trends in creative customizing, for here is a van that embodies more metalworking than meets the casual eye. More importantly, some clever innovating has been carried out; unusual treatments either entirely new or seldom seen. ■ This van's beginnings as a working shop truck found it running mundane errands for brothers Jim and Sam Radoff's custom shop with the name of Sam's Paint Shack in Detroit, Mich. It was a '71 Dodge Tradesman, 6-cyl. powered. One day when things were slow at the shop, the brothers decided to refurbished their now-worn hauler, but not alone with painting techniques, with which they were well-versed, but with a host of metal modifications. ■ The top chop is the first thing that catches the eye; a unique chop since it was lowered on an extreme wedge. Measured on the slant of the windshield posts, 14 ins. were removed, but less than half this dimension was cut from the rear. As with all top chops, the roof would have had to be either stretched to allow proper alignment between the cut-off windshield post stubs and the rear of the body, or the top could be simply moved forward in its entirely and the rear body reshaped accordingly. *Please turn to page 89*

Shown in profile at top, the wedge-shaped chop of the top is evident as is the rake to the rear of the body. Just above, radical 14-in. (on the slant) chop up front gives mail-slot appearance to the windshield. Not evident on a cursory once-over, the bobbed doors, at right, required painstaking work. The usual footwells are eliminated, and are not really needed after severe lowering of the chassis. The Radoffs have a '71 Dodge like no other.

King Kustom

Continued from page 82

stamp, just as they all carried the Barris Kustom club plaque and, later, the famed Barris crest. No matter what the customer requested, Barris managed to end up with a treatment that could be taken for no other shop's treatment. ■ When Barris Kustoms was founded in the late Forties, Sam and George were newly arrived in Southern California from Sacremento where both had been tutored under the guiding hand of Harry Westergard, one of the earliest of customizing's vanguards. At the time, customizing was surreptitiously performed either at agency body shops or in ill-equipped backyard garages for the hot rod set. But the brothers Barris brought professionalism, status and respectability to the hobby, and followed with innovations and design themes that would set standards for others to follow. ■ A Barris trademark was top chopping, and dozens of '41 - '48 Fords and Mercurys, as well as other brands, rolled out of the shop in a prodigious array. But the hallmark came with introduction of the 1949 Mercurys when Sam, in late 1948, chopped his personal 2-door that graced the cover of Motor Trend Magazine. Mercury enthusiasts flocked to Barris Kustoms, as well as to other shops within their geographical reach, for the same type of job. ■ After Sam's passing in the mid-Sixties, George carried on alone, his momentum carrying him and Barris Kustoms to the high strata of the entertainment world where cars were styled distinctively for the greats and near-greats of stage and screen. It was but a short jump from there to the creation of cars for the movie and TV studios for use in films and commercials—with the Batmobile undoubtedly the best example—and a position that George Barris continues to fill today.

"Mail-slot-windows" chop job was a Barris trademark, with Dan Landon's radical '49 Chevrolet a good example.

Valley's Customs

Continued from page 83

perhaps none emerged to become as famous as the two illustrated here; the 1950 Oldsmobile 88 hardtop of Jack Stewart, and the 1949 Ford Tudor of Ron Dunn. Valley Custom is best-remembered for its sectioning jobs, and these two examples best portray Neil and Clay's talents. It is interesting to note that the pair of craftsmen let the original styling of both cars do their "work" for them; the cars retain their basic looks and are immediately recognizable for what they are. Yet, besides the sectioning, many almost unnoticed refinements are included on each. The Oldsmobile, for example, retains its general grille shell outline, but the bumper now fronts the opening, its outer tips almost machine-fit into molded surround lines. Nearly inconspicuous vents, backed with chromed expanded metal, front each rear fender, a theme duplicated in the headlight housings. ■ Dunn's Ford retains its body side trim, though it is interrupted by the rear wheel cutouts that strengthen the car's apparent low stance. Chrome blades atop the rear fenders stream downward to halve each taillight assembly. Best of all, Valley Custom has let the metalwork do the talking; both cars utilizing single-tone paint jobs free of garishness. Both cars were built in 1952, over a quarter-century has not dimmed their greatness. The great Valley Custom shop closed its doors as 1960 neared, Neil Emory and Clay Jensen going their separate ways into more lucrative pastimes. For a time Clay did part-time collision repairs on exotic foreign cars, while Neil became bodyshop foreman for an import car dealership. Both have said they won't return to customizing as an avocation, but who is to say what the future holds?

Clayton Jensen cuts through welded, one-piece bar stock forming grille opening on the Polynesian.

Dream Truck

Continued from page 84

unassuming of starting points. During the four years it underwent transformation, each step in the process was shown in R&C in how-to-do-it detail for the benefit of custom enthusiasts everywhere. As it passed from one expert to another, each in turn was fully aware that his peers could study his techniques and handiwork. It was feared by many that any harmony of styling could not possibly result from this manner of construction, especially since each shop was allowed a free hand in its work and could easily undo—for spite—what the predecessor had done. But, in the end prejudices were overcome and as a group the specialists had torn down their self-inflicted barriers and created a harmonious, well-blended entity; a vehicle that went on to not only gain great fame but played a major role in a thesis on car customizing as an art form that earned its writer a doctorate in Fine Arts. ■ Alas, the Dream Truck, as it was known, was wrecked in a towing accident in 1958 after playing before 1½ million show attendees from Connecticut to Hawaii during the previous season. Its remains have passed through several hands since, at least two of them beginning serious restoration work before giving it up as a too-difficult cause, but its whereabouts now are unknown. ■ The Dream Truck did much to further the cause of customizing in the mid-Fifties before overly-radical custom work began reducing the prestige of automotive handcrafting to depths from which it has only recently recovered; the hobby at the outset of the Sixties spiraling down through outrageous garishness to the gloom of angelhair-bedecked "sleds" of which few were even capable of self-propulsion.

The Dream Truck was shown at most of its car shows in its white garb, so is best remembered looking like this.

Proto 1

Continued from page 85

Intended as a mid-engined car, the changing of the Valkyrie's layout became an exercise in frustration. Across town, meanwhile, Gary Wilson who, at the time, was involved with Tony Nancy, Tommy Ivo, and others of drag racing ilk, began assembling the above-described chassis as a Vette-and Porsche Turbo Carrera-eater and he needed a swoopy coupe body to mount on it. Somehow, Johnson and Wilson crossed paths, consumated a deal and the glass- and Bondo-patched Valkyrie shell, was trucked over to customizer Jerry Brown at North Hollywood Auto Body to see if he could save it. Jerry, it turns out, not only "saved" the once-Valkyrie, but added a few styling touches of his own, then used the original to pull hand layed-up panels of fiberglass. The pieces, then, were artfully united, the surfaces painted, and the whole was united with the completed Wilson chassis. ■ Though Gary Wilson now had the car of his dreams, his investment in it had to be reverted back to cash, so Proto 1 became the property of Tony Chirico, its owner at the time these photographs were made, but it has changed hands again since then. ■ Parts-spotters will be quick to identify many of the small proprietary parts that the builder(s) garnered from other cars. Doorhandles, for starters, are '69 Pontiac. Rear glass and louver panel are Porsche. Door glass is VW Karmann Ghia. Turn lights are Pontiac Firebird, and the fuel filler assembly is of Chrysler origin. Hats off to all and sundry who had their hands in Proto 1, it'll be a tough act for any builder to follow.

One-piece front end swings up for access to the 350 V-8. Car has out-run Porsche Carreras at Riverside track.

Evel Eagle

Continued from page 86

detail and an eye for cleanliness of line and form.

Most admirers comment that the Eagle is what a Datsun race car should look like, but ours is a show-and-street animal. Not to imply that this is just a promotional looker, though. The new Chevy V-8 being installed for the Oakland Roadster Show, the first public appearance of the Evel Eagle, should make it a performance winner, too. Until now, the "mean" look of the car has been enough to keep even the experienced street freaks from tangling with it. Of course, assisting this total image in the handling department has been the work performed by Dave Megugorac of The Z Place. The suspension was lowered front and rear, and stiffened to keep those pretty Goodrich T/A Radials from dicing with the fender fiberglass. Those new, wide Western Superlite wheels don't hurt the image either, they add looks and handling. Such a project never seems to be completed. What you see here is the result of only a little over three weeks of work (thrashing, really), and there are ever-newer directions CGE would like to take the project. A new grille made of body-colored louvers is yet to be installed, plexiglass headlight covers are being molded, and other changes are forthcoming, Perhaps we'll soon see duplicates of the Eagle from Datsun, either as racers or special street versions. ■ It's unlikely that Datsun would offer the chopped top or Chevy engine, though! Until such time as you can buy one for yourself, you can view this land-based Eagle in the 1978 show circuit put on by Starbird Productions of Wichita, Kansas.

Next time you see an Evel Knievel jump on TV, you may see him with this car. Not for jumping, the CGE Z project was built from a wreck for both shows and high-performance touring.

Wedged Van

Continued from page 87

The latter solution was the final choice, saving a lot of metalwork but resulting in an all-new look to the rear end. ■ The van's rear door was reshaped to conform to the extreme forward rake above the beltline, then it was mounted to hinge upward, rather to open at the side. Hinges and struts from a hatchback sedan were used to allow the lift-up action. ■ Another novel trick was the elimination of the driver and passenger door footwells. Twelve inches were bobbed from the doors' bottoms, the severed sections welded to the body and rocker panels, and new sills were formed from sheet stock. The van is lowered so radically on its suspension that the runningboard-like step that the footwells provided is no longer necessary. It's a nice but often unappreciated touch. ■ Impossible to show via the camera is the full bellypan that covers the underside of the body. It is of aluminum and, like the roof-mounted spoiler, was hammered out by Al Bargler, also of Detroit. ■ Since the Radoff Brothers are custom painters, they plied their own talents on the van's exterior; all candy colors of old-style nitro-cellulose over a gold Metalflake base. Knowing full well that too much paint is asking for later trouble, the van has only 12 coats of color and 4 of clear on it; not much by ordinary lacquer standards. ■ In 25 car and van shows, the once-lowly shop van has gathered in 25 1st place trophies, the rewards for craftsmanship and ingenuity.

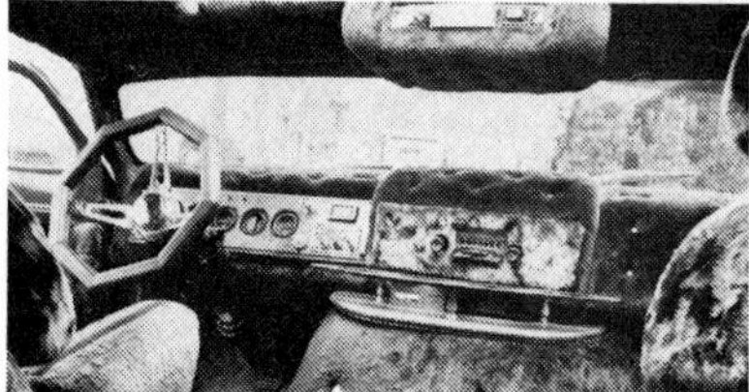

Radical windshield height reduction necessitated lowering dash 10 ins.

Standard rear door was reshaped to conform to sharp forward rake, then hinged to open like a hatchback car.

Professional Customizers and Suppliers

Creative shops, painting experts, parts and materials sources.

Following is a partial listing of professional customizing shops and specialty services, most of whom had a hand in the creativity included in this book or are otherwise noted. It is not intended that the listing include all the major professionals, for such a rundown would necessarily occupy more space than we can devote. If you wish to have custom work done rather than attempting the job yourself, and if none of the listed firms are in your immediate geographic area, consult your local Yellow Pages. Look under the heading Automobile Repairing and Painting. If there is no specific customizing listing, many large bump shops, either privately maintained or operating as part of a car dealership, will take on specialty work. It is wise, however, to study one or more examples of their work, and talk to the vehicle's owner if possible, before committing yourself.

Also included here are the major manufacturers of custom paints, tapes, airbrush spray equipment, add-on products, and other suppliers of specialty items and materials that can be useful to the professional customizer.

TAPE AND DECAL SUPPLIERS

ARLON CAL-STRIPES, INC.
19200 Laurel Park Rd.
Compton, CA 90224

G.S. STAUNTON & CO., INC.
P.O. Box 192
Birmingham, MI 48012

MRK VANS
231 John St.
Salinas, CA. 93901

NORDEC INC.
100 Northfield Rd.
Bedford, OH 44146

3M COMPANY
Public Relations
135 West 50th St.
New York, NY 10020

SPECIAL SERVICES

ARROW GLASS SERVICE
20122 Sherman Way
Canoga Park, CA

CALIFORNIA METAL SHAPING
1704 Hooper Avenue
Los Angeles, CA. 90021

DEKON ENGINEERING, LTD.
197 Peterson Rd., Box 156
Libertyville, IL. 60048

DON HARDY RACE CARS
202 Missouri
Floydada, TX. 79235

GEN-O-LITE PLASTICS CORP.
20934 Victory Blvd.
Woodland Hills, CA. 91364

JIM COOK RACING, INC.
5450 Katella Ave., Bldg. 107
Los Alimitos, CA. 90720

L & M PRECISION PRODUCTS
7750 Densmore Avenue
Van Nuys, CA. 91406

NORDSKOG COMPETITION CENTER
15917 Strathern Ave.
Van Nuys, CA. 91406

PRODUCT & SHAPE DESIGN/DEVELOPMENT/ FABRICATION
304 E. Anaheim Street
Wilmington, CA. 90744

THE FRAME SHOP
7220 Alabama Avenue
Canoga Park, CA. 91303

TRADITIONAL STREET RODS
P O Box 8186
La Crescenta, CA. 91214

Z-PLACE
19735 Sherman Way, Unit. #1
Canoga Park, CA. 91306

CUSTOMIZING AND PAINTING

BARRIS KUSTOM INDUSTRIES
10811 Riverside Drive
North Hollywood, CA. 91602

CARL GREEN ENTERPRISES
7749 Densmore Avenue #7
Van Nuys, CA. 91406

CHUCK PORTER
10541 Pleasant Way
Whittier, CA. 90670

CRAZY PAINTERS
9665 Alondra Blvd.
Bellflower, CA. 90706

CUSTOMS BY EDDIE PAUL
123 Nevada Street
El Segundo, CA.

DARRYL STARBIRD
RR #1, Box 385
Mulvane, KS.

DEAN JEFFRIES' AUTOMOTIVE STYLING
3077 No. Cahuenga Ave.
Hollywood, CA. 90068

DICK DEAN AUTO RECONSTRUCTION
6920-D Knott Avenue
Buena Park, CA. 90620

JOE BAILON
5955 Troost Avenue
North Hollywood, CA. 91607

WINFIELD SPECIAL PROJECTS
7749 Densmore Avenue, #6
Van Nuys, CA. 91406

ADD-ON PRODUCT SUPPLIERS

A & A FIBERGLASS, INC.
1534 Nabell Ave.
Atlanta, GA. 30344

AUTOMOTIVE ACCESSORIES UNLIMITED
2609 Woodland
Anaheim, CA. 92801

AUTOMOTIVE DEVELOPMENT
501 W. Maple, Unit. V
Orange, CA. 92668

CALIFORNIA COMPONENT CARS, INC.
2333 Grant Ave.
San Lorenzo, CA. 94580

CALIFORNIA STEP SIDE, INC.
P O Box 60249
Sunnyvale, CA. 94088

DICK BARBOUR DATSUN
5800 Lincoln Ave.
Cypress, CA. 90630

DOMUS
312 E. 17th St.
Kansas City, MO. 64108

FANTASY FACTORY
12545 Sherman Way
N. Hollywood, CA 91605

FIBERFAB, INC.
548 Baldwin St.
Bridgeville, PA. 15107

HICKEY ENTERPRISES, INC.
1645 Callens Rd.
Ventura, CA. 90630

KARVAN INTERNATIONAL, INC.
1401 S. Village Way
Santa Monica, CA. 92705

MAIER RACING, INC.
235 Laurel Ave.
Hayward, CA. 94541

MANSFIELD FIBERGLASS & PLASTIC CO.
P O Box 127
Thorntown, IN. 46071

SATYR ENTERPRISES, INC.
15153 Garfiend Ave.
Paramount, CA 90723

SED PRODUCTS
1158 North Gilbert St.
Anaheim, CA 92801

SPARTAN PLASTICS INC.
P.O. Box 67
Holt, MI 48842

WIKE FIBERGLASS
2117 Palma Dr.
Ventura, CA. 93003

PAINTING EQUIPMENT

BADGER AIRBRUSH CO.
9201 Gage Ave.
Franklin Park, IL 60131

BEUGLER STRIPING TOOLS
P.O. Box 29068
Los Angeles, CA 90027

BINKS MANUFACTURING CO.
9201 W. Belmont Ave.
Franklin Park, IL 60131

DE VILBISS SPRAYGUNS
P.O. Box 913
Toledo, OH 43692

K.J. MILLER AIRBRUSHES
2401 Gardener Road
Broadview, IL 60153

PAASCHE AIRBRUSH CO.
1909 Diversey Parkway
Chicago, IL 60614

STEWART-WARNER CORP.
1826 Diversey Parkway
Chicago, IL 60614

THAYER & CHANDLER
442 North Wells St.
Chicago, IL 60610

PAINT PRODUCT MANUFACTURERS

AERO-LAC AUTOMOTIVE FINISHES
SEM Products, Inc.
Sem Lane & Shoreway Rd.
Belmont, CA. 94002

DITZLER AUTOMOTIVE FINISHES
PPG Industries, Inc.
P O Box 5090
Southfield, MI. 48037

E.I. Du PONT DE NEMOURS CO.
10th and Market Streets
Wilmington, DE. 19898

MEARL CORPORATION
41 East 42nd Street
New York, N.Y. 10017

METALFLAKE, INC.
Haverhill, MA. 01830

R-M PRODUCTS
Inmont Corp.
5935 Milford Ave.
Detroit, MI. 48210

SEELIG'S CUSTOM FINISHES
P.O. Box 179
Inglewood, CA 90306

SHERWIN-WILLIAMS
Automotive Finishes Div.
4133 Payne Ave.
Cleveland, OH 44103

TIVIAN LABORATORIES, INC.
P O Box 6448
Providence, RI. 02904

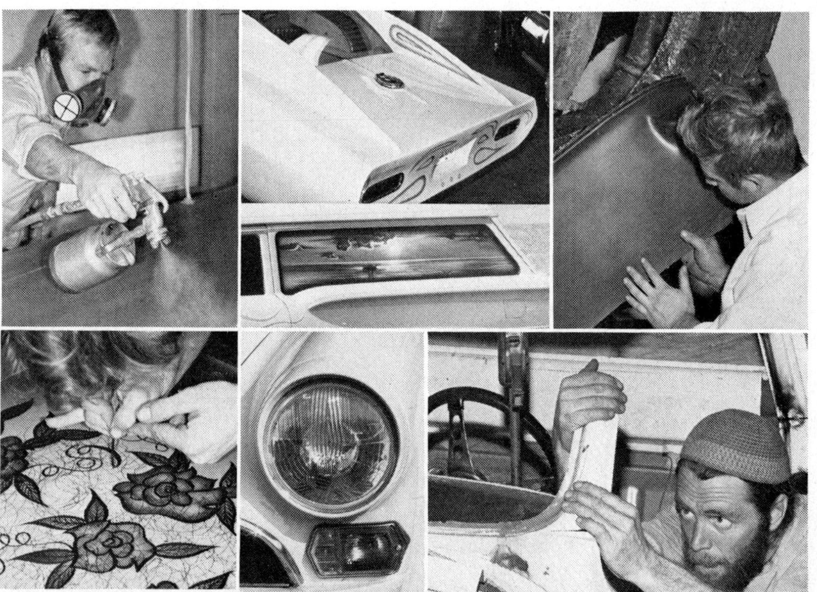

A Colorful History

From the "only black" of yesteryear to the glittering custom finishes of today.

by Richard Busenkell

PHOTOS COURTESY THE RESPECTIVE MANUFACTURERS AND PPC LIBRARY.

The history of customizing is really the intertwined history of two different but related arts—bodymaking (coachwork) and painting. We dealt with coachbuilding in an earlier section, so now let's look at the history of paint and its application to automobiles.

Paint is as old as man. Cave drawings many thousands of years old are in color. By the time of the ancient Egyptians, history's first great culture, painting was already developed into a high art; witness the brilliant colors on the fabled treasures found in Tutankhamen's tomb. On objects not so carefully protected, however, paint did not last so long; just as today, paint exposed to weathering rather soon faded. It is not generally realized that the great marble portraits and statuary of the Greeks and Romans were painted; only the tiniest traces of that paint remain today.

ENAMEL

Any paint is composed basically of two parts. The vehicle is the liquid which allows the paint to be flowed or sprayed, and it must form a hard film upon drying. The pigment is the material which gives color to the paint. Either vehicle or pigment can be a single substance or a mixture of several materials. Pigments can be liquids, but they are generally a finely ground powder which is held in suspension by the vehicle. For the color to be uniform, the pigment must be evenly dispersed throughout the vehicle; that's why stirring or agitating paint while using it is so important.

The first paints used in automotive work were derived from natural products, especially two: the flax plant and the pine tree. Flax was already a most useful plant since its fibers, when spun and woven, make linen. Its seeds, however, when crushed, yield linseed oil, which is still today a popular paint vehicle in Europe and America. Raw linseed oil dries very slowly and is used primarily in metal-protecting paint, such as bridge paint; boiled linseed oil dries more rapidly, and is used as a wood preservative as well as a common paint vehicle. When boiled, the sap of a pine tree separates into rosin, that substance so beloved by violinists and baseball pitchers, and a vapor, which when collected and cooled into a liquid is called turpentine. Now; if rosin and boiled linseed oil are mixed together, the result is varnish. If pigment is added, the result is color varnish, or enamel. For varnish and enamel the thinner is turpentine, which not only cuts rosin but was found to satisfactorily dilute linseed oil.

EARLY PAINTING PROBLEMS

Painting an automobile body in the old days with varnish or enamel was a long and painstaking task. Note that we said "automobile body," not "car," since bodies were painted separately before being installed on the chassis. There was really no customizing as we know it today, since all bodies were custom built. Then too, almost all accessories were the province of the dealer, not the auto manufacturer. When you walked into your Packard, Pierce-Arrow or Ford showroom, you could select from a vast array of aftermarket goodies which the dealer would install on your car after it arrived from the factory.

Varnish and enamel took a long time to dry, because linseed oil takes a long time to dry. For any manufacturer with any hopes of high production, the body finishing area took up an enormous amount of space. It was found that black enamel, in which plain carbon black was the pigment, was the fastest drying of all enamels. It was for this reason that Henry Ford, the greatest volume producer of his time, insisted that a Model T customer could have "any color he wanted, as long as it was black."

1

2

3

4

5

1. Fisher bodies being built in 1918 show the type of handwork necessary in the days of metal over a wood frame.

2. After assembly and priming, early auto bodies were rubbed with felt pads and crushed pumice, usually by apprentices. A smooth flat surface was essential since there was no rubout of the final varnish coats.

3. The final varnishing was brushed on by skilled and patient painters. The result was spectacular, but took 10 to 12 days even at fast-moving GM and didn't last very long.

4. Paint-spraying equipment came into use during the Twenties. Walls washed with running water kept the air clean when spraying large body pieces, such as these 1937 De Soto fenders.

5. Hudson bodies move down a Detroit assembly line in 1930 as girls pinstripe the bodies with a device little remembered today, the striping machine. By this time, most auto manufacturers were using Duco lacquer.

Any other color would have held up his production. Even so, painting a Model T still took three days; but for a fancy Pierce that had several different colors, the painting could take up to three months! Yet despite all this labor, the average paint job did not last very long, unless it was very carefully protected indeed; about three years was the average, unless the car was always garaged when not in use, never left out in the sun, and never driven in the rain.

BAKED ENAMEL

Around 1915 came two developments which had a strong impact on auto painting. The first was the development by the chemical industry of a synthetic rosin which could replace the pine tree product in enamel. To distinguish this synthetic material from the original, and to avoid lawsuits, it was spelled resin, which is now the term for all synthetic and natural substances which resemble rosin. This resin required a high temperature to dry, but allowed a much shorter drying time—about one hour. It was found that carbon black was the only pigment really stable at the elevated drying temperatures required by this resin, so black was the only color to be offered until 1923. This was no real problem, since black was the basic car color anyway in those days, and the paint became known as black baked enamel. Besides its shorter drying time, this paint was much more durable than ordinary enamel, retained its shine, needed little polishing, did not crack or check, and resisted weathering very well. So great were the advantages of this paint that auto companies did not hesitate very long before investing in the gas-heated drying booths necessary.

There remained a problem, however. Normal auto construction was metal over a wood frame, and that wood could not take the high drying temperatures. Therefore the usual parts finished in black baked enamel were hoods and fenders, which had no wood framing and were painted before assembly; the rest of the car was painted in ordinary enamel.

The second development offered an answer to this problem. In 1912 the Budd Company, then a major auto body supplier and more recently a principal builder of railroad cars, manufactured and displayed several all-steel car bodies. John and Horace Dodge, who had made a fortune supplying engines and components to Henry Ford, brought out their own car in 1914, and quickly adopted the all-steel Budd technique. The 1916 Dodge tourer was history's first all-steel volume-produced car, and it was finished in black baked enamel. Naturally, this required the construction of huge gas-fired heated drying rooms, an immense capital investment for a new company. The gamble paid off, however, as the 1916 Dodge shot up to become the fourth best-selling car in America that year. Despite this success, all-steel construction did not become commonplace among American cars until the second half of the '30's. Baked enamel is still very much with us, and remains today the most popular paint finish, though in somewhat different formulation from that of 60 years ago.

DUCO

In 1922 the E.I. DuPont de Nemours & Co.—the Delaware-based chemical giant already famous for its explosives and destined for immortality as the future inventor of nylon, ®Orlon, Dacron, and Teflon,—introduced a totally new paint system which found almost immediate application in the auto industry. It did not dry by oxidation, but rather by the rapid evaporation of volatile solvents in its vehicle, leaving a nitrocellulose film on the metal surface. It was compatible with a wide variety of pigments, and since the drying process could be done quickly at room temperature, these pigments did not need to be stable at elevated temperatures, meaning a great variety of colors were instantly available. Light in weight and thin in viscosity, it could be easily sprayed; since it dried quickly, mistakes such as runs could be easily corrected. DuPont called this new paint Duco, and it took the auto industry by storm.

Interestingly, DuPont did not call this new nitrocellulose paint "lacquer," at least not at first; it was simply Duco. However, DuPont allowed other paint manufacturers to make Duco under license, and they noted that when polished, Duco had a gloss rather resembling that of antique Japanese and Chinese furniture and decorative wooden items. This finish, called "lacquer" by Europeans, thus gave its name to Duco, and by the early '30's lacquer was the common name for nitrocellulose paint. This practice was doubtless encouraged by rival paint manufacturers, who were not anxious to have the public demand a paint which by its very name was a reminder of DuPont. The joker here is that "lacquer" was a term which was technically not correct even for the Japanese and Chinese wood finishes, which were derived from the sap of an Asiatic tree (Rhus vernicifera). This finish was called lacquer by non-Orientals of the 19th century because it resembled a wooden surface treated with shellac, or as it is known in much of the world, goma lacca. Commercial shellac is the solution in alcohol of the sticky resinous secretion left on tree limbs by the tiny lac insect, Laccifer lacca, which is widespread throughout India and southeast Asia. Thus the paint we call lacquer today is actually twice removed from the origin of its name.

Another interesting point about Duco is that initially it was not polished. Enamels do not need polishing after they have hardened; therefore it was felt to be a black mark against this new paint if it required a lot of hand labor after it

1

2

3

4

1. GM has long used lacquer, which means plenty of handwork even in a highly automated factory. These 1958 Fisher "bodies-in-white" are given a rough rubout before the final coat.

2. Hand spraying was still the best way to paint a car in the Fifties, although protection for the painter had improved. This is a '59 Imperial.

3. Still on the same conveyor line which took them through the baking oven, GM cars are given a final buffing before being delivered to dealers.

4. Like all GM cars, this 1958 Pontiac is led for a 7-minute stroll through an infrared drying oven after final trim assembly. This short baking adds hardness to the lacquer finish.

was applied. You know today what unpolished lacquer looks like—it has a satiny sheen, but is not glossy. That is exactly what DuPont tried to sell the auto industry in 1923; the paint was even called Satin Finish Duco.

DUCO TAKES OVER

The first car to be factory-finished in Duco was the 1924 Oakland, GM's medium-priced car which the Pontiac would later replace. Blue was the only color offered, which Oakland called "True Blue." By the end of the following year the entire GM line was in Duco, as well as Chrysler and a host of other respected makes, such as Franklin, Chandler, Nash, Marmon, Kissel, Maxwell, Hupmobile, and even the American Rolls-Royce. In 1926 many of these cars appeared in polished Duco, which entailed additional work but was worth the effort; DuPont stopped talking about satin finish. In 1927 the prestigious makes of Peerless and Pierce-Arrow joined the Duco fold, as well as enclosed-body Fords. The next year all Fords had Duco, a situation which sorely vexed Henry Ford, since GM owned 39% of Du-Pont stock! (This near-monopolistic control of a major auto supplier was corrected by the federal government in 1935, when the Justice Department forced GM to divest itself of DuPont stock.)

The widespread use of Duco forever ended Henry Ford's famous "any-color-as-long-as-it's-black" days. A riot of color spread through the low-priced cars. Chevrolet was using Duco in 1926, one year before Ford's enclosed cars and two years before Ford's open cars. This was certainly a factor in Chevrolet's outselling of Ford in 1927 and 1928, although the principal reason for Chevrolet's success was the shutting down of Ford production in 1927 while tooling for the new Model A was installed.

Duco was also a boon to custom coachbuilding. Although structural wood was disappearing from mass-produced cars, it was still extensively used in luxury cars, since they had custom coachbuilt bodies which were still made the traditional way, metal over a wood framing. The great coachbuilders which are so much a part of American automotive folklore—Brunn, Rollston, Murphy, Le Baron et al—were really very small operations compared to the auto manufacturers, and the ability to use a paint which did not require large drying rooms was a great advantage. A greater variety of color was demanded by customers of coach-built cars, and Duco filled that bill admirably. All expensive cars of the late '20's and '30's were painted in nitrocellulose lacquer, either Duco or its imitators. So pervasive has been the influence of this paint that even today in modern-day Britain the common name for all automotive paint is "cellulose," even if the particular paint referred to is enamel!

ENAMEL STRIKES BACK

Henry Ford was not pleased with the fact that his own cars had to be painted with lacquer. It would be easy enough to state that this was because archrival GM owned so much of DuPont, but there was a lot more to it than that. Although the industry-wide switch to lacquer eliminated the need for expensive drying rooms, lacquer had its own problems. Because a finished lacquer paint job, even with many coats, was only about half the thickness of an enamel job, grinder marks, sand scratches and other surface imperfections showed through clearly which would have been hidden by enamel. So the quality of prepainting metal treatment had to be significantly upgraded, calling for a higher number of skilled workmen. Then after the painting, yet more skilled labor and valuable production space was needed to carry out the color sanding, rubbing and polishing operations. So the search for the best automotive paint was far from over.

If the term had not already been applied to William Jennings Bryan, Henry Ford would certainly been called the "Great Commoner." He designed his cars for the low-priced market, the common man, and forever sought ways to involve the agricultural heartland of America in the riches of the automotive revolution. There is a famous photo of Ford swinging an axe at the trunklid of a Ford sedan to demonstrate the toughness of a soybean-based plastic from which the experimental trunklid was made. Such plastics never found much use on mass-produced Fords, but a new enamel with soy oil replacing the traditional linseed oil did. Like linseed oil and certain other oils sometimes used as enamel bases (tung oil and dehydrated castor oil, to name two), soy oil is a vegetable oil which dries. (Mineral oils, incidentally, are never used in enamel since they do not form a film.) Soy oil is lighter in color than linseed oil, and resulted in brighter colors as well as harder paint; linseed oil became relegated to primers. Since they used synthetic resins, these new enamels became known as synthetic enamels, though that term might just as well have been used for previous enamels with synthetic resins. The year 1938 saw the beginnings of changeover from lacquer to synthetic enamel at Ford, and Chrysler followed not long after. To this day Ford and Chrysler use enamel while GM sticks with lacquer, though neither

1

2

paint system is the same now as it was then.

The next breakthrough in paint was the development of alkyd enamel, in which a synthetic alkyd resin was allied with an oil base. For automotive use this base was usually coconut oil, which was unlike previous oil bases in that it never really dries. This use of a non-drying oil imparted a greater flexibility and resistance to cracking to the paint. Ford made the shift to this type of paint about the same time that GM was turning to acrylic lacquer, during the span 1955-57.

MODERN SYNTHETICS

The miracles of the modern plastics industry have changed today's automotive paints, even though they are still clearly divided into enamels and lacquers, with some new types added. Synthetic resins have long since replaced the rosin and even the oils in enamels and nitrocellulose in lacquers.

There are a rather staggering number of synthetic resins now in use, and we certainly can't go into them all here, but chemists divide them into two basic types: condensation polymers, which include polyesters, phenolics, amino resins, alkyds, epoxies, polyurethanes, and addition polymers, which include polyethlenes, polyvinyl acetate,

4

3

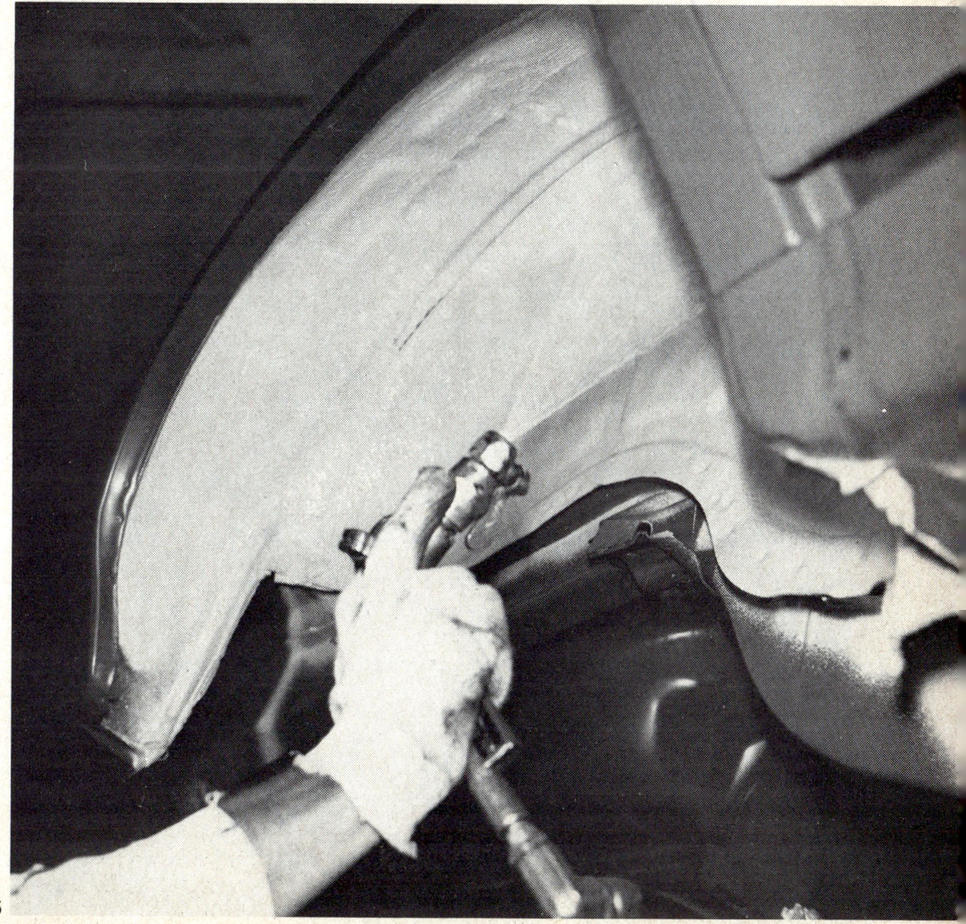

5

1. After primer, modern Fisher bodies are hand-sprayed with finish color on edges and body crevices before entering automated spray booth.

2. New GM cars now have primer applied a different way. Fisher body for a 1977 Chevrolet takes a dunking in electrophoretic primer bath, in which the body and paint have opposite electrical charges. Primer seeks out every nook and cranny this way.

3. Detroit pinstriping is done today in almost the same manner as it was 50 years ago, except that a metal guide is used. Lowered around the entire car, this guide serves as a template for the stripe applicator.

4. In the automatic spray booth, paint nozzles situated all around the car cover the entire exterior surface. Color sequence of cars on the paint line is governed by computer.

5. Nobody has yet figured out a way to automate the application of undercoating, so that's one job that's still done by hand-held spray gun.

polyvinyl chlorides, and acrylics. All of these have their specific uses, but some of them have an enormously widespread use. The most common type of house paint for example, is alkyd enamel, in which an alkyd resin (there are several different ones) mixed with linseed oil (or other oil) is the vehicle; over one billion pounds of this type of paint alone are produced annually. The type of enamel you buy in a spray can is also alkyd enamel, unless the label specifically states otherwise. Latex paints are aqueous emulsions of polyvinyl resins, and are soluble in water when wet; DuPont's Lucite is a famous latex brand. Latex paints are usually called "latex enamels," though this seems to violate the general definition of enamels as oil-based paints.

Of chief importance to present-day automotive finishes are the acrylates. These acrylic compounds, which are clear, non-yellowing and stable at high temperatures, are the basis for all modern automotive paints. Acrylic enamels also have no oil, and are made (brace yourself) by co-reacting acrylic monomers with another synthetic resin, usually styrene, to form a co-polymer, which is then mixed with melamine formaldehyde; the molecules in this mixture cross-link when baked, forming a tough film. That's the stuff that's on the outside of all new Ford and Chrysler cars. GM, a lacquer lover since the days of good old Duco, has been using acrylic lacquer for two decades. The changeover from nitrocellulose lacquer took place during the '56-'57 model years, and the entire GM line was in acrylic lacquer by 1958.

TWO-PART PAINTS

Epoxies, polyurethanes, and two-part enamels are the latest paint developments. Though these three are different paint systems, they are similar in that they all use a separately added catalyst to cause the chemical reaction which hardens the paint; they do not dry by oxidation or evaporation (except the two-part enamels, in which the catalyst speeds up an air-started reaction). Epoxies do not have the resistance to fading of the later-developed polyurethanes, which are finding great favor among restorers as coatings for chassis and underbody parts since they are so resistant to chip damage.

The real up-and-coming paints appear to be the two-part enamels. In paints such as Ditzler's Delstar/Delthane and DuPont's Centari, a catalyst added to the enamel causes it to set up almost as fast as lacquer. Dry to the touch in just a few minutes, these paints can be left as is, for they dry glossy like any enamel. They flow on smoothly and hide sand scratches well, again like any enamel. But they dry so fast and so hard that they can be color-sanded to a flat brilliant gloss within 24 hours of application, which is unheard of for enamels. Since the catalyst kicks off the chemical reaction, there is no need to bake these paints, eliminating another traditional objection to enamels and making them very suitable for the amateur or customizer with only a small shop at his disposal. In fact, so revolutionary and different are two-part enamels that it is best to think of them as an entirely new paint system, which combines the best qualities of previous enamels and lacquers.

Two-part enamels should not be confused with two-stage enamels, or the basecoat/clearcoat treatment as it is sometimes called. Already an option on several Ford Motor Co. cars ranging from the new Ford Fiesta to the Continental Mark V, this paint treatment gives a shiny "wet" look without the necessity for rubbing.

If you get the impression from all this that chemists are the most im-

1

portant people at modern paint plants, you're right. The automotive paint industry has gone through several revolutionary changes since the days of hand-brushed varnish jobs, and the search for the perfect paint is far from over. There will probably be no more reason for the cars made in the year 2050 to be using today's acrylic lacquers and acrylic enamels than there is for today's cars to have those varnish jobs of 1910.

CUSTOMIZING

What does all this paint history mean to you, aside from the intrinsic interest of it if you're a customizer? Well, for one thing, since customizers often apply their art to older vehicles, it's a help to know what the original paint was if a complete stripping is not feasible. Most customizing, for example, is still done in lacquer, and you cannot just go blindly shooting lacquer over the previous paint and expect that everything will come out perfectly. The old rule of thumb is still a good one: you can paint enamel over lacquer, but you can't paint lacquer over enamel. This is (or was) especially true in the days of alkyd enamels and nitrocellulose lacquers, as the flexible enamel would crack the brittle lacquer sprayed on top of it. Incompatibility of the two paint systems is not nearly so much of a problem with modern acrylics, but it is still a point which cannot be ignored.

Customizing, as we understand the term today, is the art of changing a stock automobile into a personalized creation. Leaving aside any mechanical changes, a custom automobile can be created by changes to the bodywork, or a new paint job using non-original paints, colors, and techniques, or a combination of both. Such changes were not common before World War II, for two reasons. The first was that there were a goodly number of custom coachbuilders who would build a whole custom

2

3

4

5

1. Here's the reason people customize. Acres of lookalike cars—Chevettes here—pour out of Detroit by the millions per year. Anyone who wants true individuality has to be creative.

2. Dean Jeffries started with George Barris, then opened his own custom shop. Famed as a striper, he sprang to prominence as a customizer with his "Manta Ray," based on a Grand Prix Maserati. He now makes special vehicles for the movie studios.

3. The first pearl-painted custom was the Golden Sahara, displayed in late 1954. Barris rendered the original '54 Lincoln almost unrecognizable, stretching it and adding a TV set and a complete bar.

4. This chopped '54 Oldsmobile won a second place in its class at the 1961 National Custom Car Show. It's an excellent example of pearl paint.

5. The Kopper Kart was a famous show vehicle of the late Fifties. Built by Barris in 1956 from a Chevrolet pickup, the Kart was the first custom to use diamond dust paint. Mixed with pearl, this lustrous and very expensive paint was called "diamond pearl."

body to suit the taste of any buyer, so long as he had money. And the not-so-wealthy could really "customize" their car at their local dealer, since factories tended to ship their cars stripped and leave accessories to be installed at dealer expense.

Following the Depression, World War II, the demise of the classic-era coachbuilders, and the increasing tendency of the auto manufacturers to fit accessories at the factory, about the only way anybody could obtain a custom American automobile was to customize a stock car.

Customizing with paint can be divided into two categories: the custom paints themselves, and techniques used in applying them. Let's take a look at the history of both.

CUSTOM PAINTS

Plain nitrocellulose lacquer was something pretty special in the Twenties and Thirties, to a generation of Americans just growing out of the all-black Model T. But there are always those who want something even more, and metallic paints were the first truly custom paint to appear. Consisting of a powdered metal (usually aluminum) suspended in the vehicle and kept dispersed by constant stirring, metallics appeared on a few custom-bodied cars in the late Thirties; the 1940 Packard line catalogued several metallics as optional finishes on its higher-priced models. There were some problems with early metallics. Aluminum oxidizes rapidly, turning a dull gray, so the best procedure was to add the metal to the paint immediately after turning it into a powder, which was often not possible.

The first custom paint of the Fifties was candy apple, so named because of its brilliant red color. A silver metallic basecoat was overlaid by a transparent red coat, formed by adding red toner to clear lacquer; this was followed optionally by a clear topcoat. The exact origin of candy paint is disputed; credit for it is claimed by Joe Bailon, George Barris, and Emmet Glasgow, who worked for Barris at the time.

The first Barris candy job was a chopped '41 Buick on display at the first Motorama, held at the Los Angeles Armory in January 1948. George says his inspiration came from some Christmas tree ornaments that he observed in the winter of 1947. Locating a Los Angeles manufacturer of these ornaments, he learned that the brilliant red color came from the addition of a red dye to clear lacquer, which was then sprayed over the silvery surface. After considerable experimentation, he then attempted a similar process with red-toned lacquer over over the chopped Buick, which was first painted with silver metallic. The color came out very dark, for two reasons; George had added too much toner to the clear, and the powder particle size in the metallic was too fine to reflect much light through the topcoat. Only later did paint manufacturers realize that the only way to get a metallic paint to really shine was to increase the particle size.

Joe Bailon's first candy job was his '41 Chevy exhibited at the Oakland Roadster Show in February 1952. He remembers that he got his idea from examining the highly reflective paint on the drums of a famous swing-era drummer.

Like the automobile itself, it appears that candy paint was developed by several people working independently.

Whatever its origin, candy paint caught on rapidly. Bailon and Barris were soon offering versions of it in spray cans. Colors other than red soon followed, and of course they could be sprayed over any base metallic color.

Pearl or pearlescent paint appeared not long after candy. Like candy, pearl is shot over a base color and followed by a clear topcoat. The first pearl-painted show car was the Golden Sahara, another Barris creation first shown in 1956. Basically a '54 Lincoln, the Golden Sahara was so modified that the original car was unrecognizable; it included a TV set, telephone, tape recorder, and bar equipment. Barris first saw the pearlescent paint on the signboards of an advertising firm. Inquiring, he found that the effect was produced by ground fish scales, and that the special pigment containing them was available only in Japan. Finally obtaining some, he then painted the show car. As George now tells it, the shimmering golden color was an accident. After painting the pearl over a white base and sealing it with clear lacquer, the car was pushed outside to dry. Exposed to the hot California sun, the fish scales quickly turned yellow, so Barris and his crew coined the name "Golden Sahara" and gold plated the trim to match!

Disappointed with the poor durability, high cost, and near-unavailability of the fish scales, Barris then tried crushed abalone shells, which weren't much better. Modern pearls use a crushed mica, which is much more stable since it is a mineral and not an animal product.

Pearl also eventually was offered in spray can form; Dean Jeffries marketed such a product in 1962.

Metalflake came next, and before we go any further we should point out that the name "Metalflake" is a registered trademark of Metalflake Inc., a company in Haverhill, Massachusetts. Like Coca-Cola, Kleenex, Scotch Tape and Fiberglass this trade name has become so closely identified with the product that it is often used generically to denote any similar product. Competing companies were

1. Donn Varner was the striping king of the Pacific Northwest in the early Sixties. His work was marked by an imaginative style and meticulous detail.

2. Gene Winfield's "Jade Idol" was shown at the 1962 National Custom Show at Indianapolis. A customized '56 Mercury finished in varying shades of green, the Jade Idol was the first example of "fading," the blending of one shade of color into another.

3. Another custom which started as a '56 Mercury is this wildly scalloped convertible. Hailing from Michigan, this car placed second in its class at the 1959 National Custom Show.

4. This '60 Chevrolet is a good example of early-Sixties scalloping.

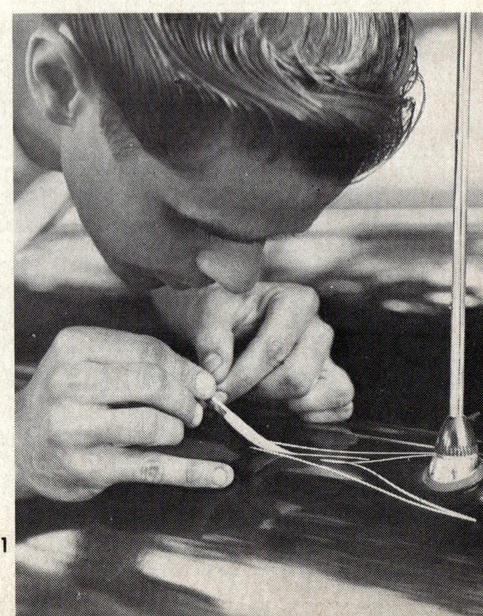
1

forced to dream up different names for this type of paint; Aero-Lac, for example, uses the term "Aero-Flake," while Ditzler calls their version "Sun-Gleam."

Metalflake is considerably newer than candy or pearl; it was introduced in 1960 by the Dobeckmun Co., a division of Dow Chemical, which is now Metalflake Inc. Other than a Chrysler painted by the Dobeckmun Co. as an advertising display, the first show car to be flaked was the Barris XPAK 400, a Martian-looking air car shown at the 1960 National Championship Custom Show in Detroit.

Actual flakes of metal are not used in these "metal flake" paints; they are tiny rectangles of mylar. The mylar can be dyed any color, but silver and gold are used almost exclusively. This type of paint has great versatility, for the size as well as the color of the tiny mylar flakes can be varied; the Metalflake Co. offers four sizes. The larger the size

2

3

4

of the flake, the greater the brilliance and reflectivity.

Truly original custom colors can be created by combining any of the above in endless variations. Pearls can be substituted for metallics as the base for candy colors, and candies can be shot over flake for dazzling colors.

In formulation, most custom paints traditionally have been lacquers; acrylic lacquer is the mainstay today. Custom painting involves several coats at a minimum, and lacquer's abilities to dry quickly between coats and bond well to a previous lacquer coat are greatly appreciated. Lacquer, of course, can also be polished to a brilliant gloss and depth, but this fact is less important than it once was; pearls, candies, and flakes get their luster from a clear topcoat, not from a laborious rubout.

A new type of paint which may threaten acrylic lacquer's popularity with customizers is acrylic urethane. This paint dries by the same process as polyurethane enamel, but due to its base acrylic, it has metallic color capability. Like polyurethane it has great resistance to weathering and solvent attack, but what really makes customizers sit up and take notice is, this paint's ability to be applied without a sealer or primer directly over acrylic or nitrocellulose lacquers, acrylic or alkyd enamels, polyurethanes, or even bare aluminum!

TECHNIQUES

The oldest custom technique is pinstriping, which goes back to the age of carriages. Delicate pinstriping was common on luxury cars during the classic era, and also found extensively on less expensive cars as the volume manufacturers sought to mimic the appearance of the coachbuilt masterpieces. Pinstriping quietly faded from the scene in the late Thirties and Forties, but staged a strong comeback in the customizing of the Fifties. There was one essential difference, however. Previous pinstriping had been done to enhance and highlight the basic body lines of the automobile. It was always understated and discreet, often not noticed at first glance. Customized pinstriping was quite the opposite; it was designed to be noticed, a dazzling art form which used the auto body as a canvas. The outlines of flames, scallops, arabesques, intricate webs, delicate lace, and fanciful geometric figures danced from the striping brushes of Dean Jeffries, Donn Varner, Larry Watson, Damon Ritchey, Tommy the Greek, Ed Roth, Andy Southard Jr., Katayanagi, and the most legendary of them all, Von Dutch. A striping job bearing the signature of Von Dutch, or even better, his personal trademark of the "flying red eyeball," was enough to send every car freak in the neighborhood into a tizzy.

From outlining flames with pinstripes, the next natural step was to fill in the flames. A righteous flame job became the next sign of a leading custom, and the new candy and pearlescent paints really made the flames come alive.

Countless were the customs created and exhibited during the heyday of customizing, which stretched roughly from 1947 to 1960. It would be pointless to even try to mention them all, since there were so many. After all, the whole idea of customizing is to make the car look different from any other.

Eventually and inevitably, the auto manufacturers themselves entered into customizing. At first this was manifested only in mechanical options such as 4-bbl. carburetors, special suspensions and heavy-duty brakes, but a few years later the factories blossomed forth with fancy paint schemes, decals, zoomy stripes, alloy wheels mounting fat tires, and all sorts of spoilers and scoops. To a great extent this satisfied the customer's need for individuality, and customizing declined.

What caused a resurgence in customizing, especially custom painting, was the rather unexpected explosion of interest in pickup trucks and vans. By their nature as utilitarian vehicles, the various brands of pickups and vans are about as much like one another as peas in a pod, offering a fertile field for customizing. And the broad sides of vans were ideal for a new fad in custom painting: murals. No one seems to know who did the first van mural, for it seems such a natural thing to do.

However, Car Craft magazine in its February 1956 issue showed a Von Dutch mural on the side of a '50 Ford station wagon; done in 1952, the wagon was used as a

1

2

delivery and parts vehicle for George Barris, and was probably the first automotive mural. The first murals were freehand, which still provides the most stunning result, but stencil murals are now popular since they are cheaper.

New paints specifically for customizing have developed, such as Eerie-Dess and Metalflake's Vreeble, but these are really variations of the four basic custom paints: metallic, candy, pearl, and flake.

Several new techniques have also risen. Cobwebbing is the spraying of unthinned lacquer out as a fine thread, moving the spray gun around in order to create an intricate spidery network. In lacing, a stencil cut like lace or an actual piece of lace (hopefully cheap) used as a stencil creates a soft delicate tracery.

CUSTOM PAINTING TODAY

We rarely see a wild "Kalifornia Kustom" in the grand manner of the Fifties today, but that is because customizing and custom painting have changed their venue. With the great proliferation of models from Detroit to satisfy almost every styling whim, the attention of customizers has shifted to trucks and vans. If Detroit picks up on this fad and brings out its own lines of customized vans and trucks (such a movement is already beginning), the custom field will turn to yet another field, perhaps minicars or off-road vehicles.

In custom bodywork, kits with flares, spoilers and special grilles have replaced many of the handcrafted one-of-a-kind customs of two decades ago. In the painting field, the equivalent of these are stencils and vinyl tape, which many custom shops use to turn out custom vehicles (especially vans) which are much like one another. Thus volume is gained with a loss of individuality.

There never will really be an end to customizing. It began as a gentle revolt against the conformity of Detroit's mass-produced vehicles, and that revolt continues. New paints and techniques will undoubtedly appear, and the emphasis may be shifted from one vehicle type to another, but these will change merely the form of customizing, not its substance. As long as Detroit is in the business of mass-producing millions of transportation machines each year, there will always be a few thousand with the taste, time, and creativity to say, "That's not for me."

1. Multi-talented Andy Southard Jr. practiced his striping skills in the San Francisco area in the late Fifties, and is still in demand. An excellent photographer known for his many fine photos of custom cars, Southard also worked for the California Department of Highways.

2. The most legendary and reclusive of stripers was (and still is) Von Dutch. He was the first to turn pinstriping from a mere accent of a car's basic styling into an imaginative art form all its own, the first customizing pinstriper. He was also the first known muralist and an early flame stylist.

3. First "car" to use Metalflake was the Barris XPAK-400, an experimental air-cushion vehicle which won top prize in Manufacturer's Division at the 1960 National Custom Show.

4. After its sensational debut on the XPAK-400, Metalflake paint became the hottest thing in custom-car circles. At car shows around the country, the booth of the Dobeckmun Co., manufacturers of Metalflake, was visited by enthusiasts seeking information about this new paint.

5. Flake, striping, swirls, bars, even a little mural work—this car has it all, and shows the great variety available to today's customizer.

3

4

5

Guide to Paint Materials

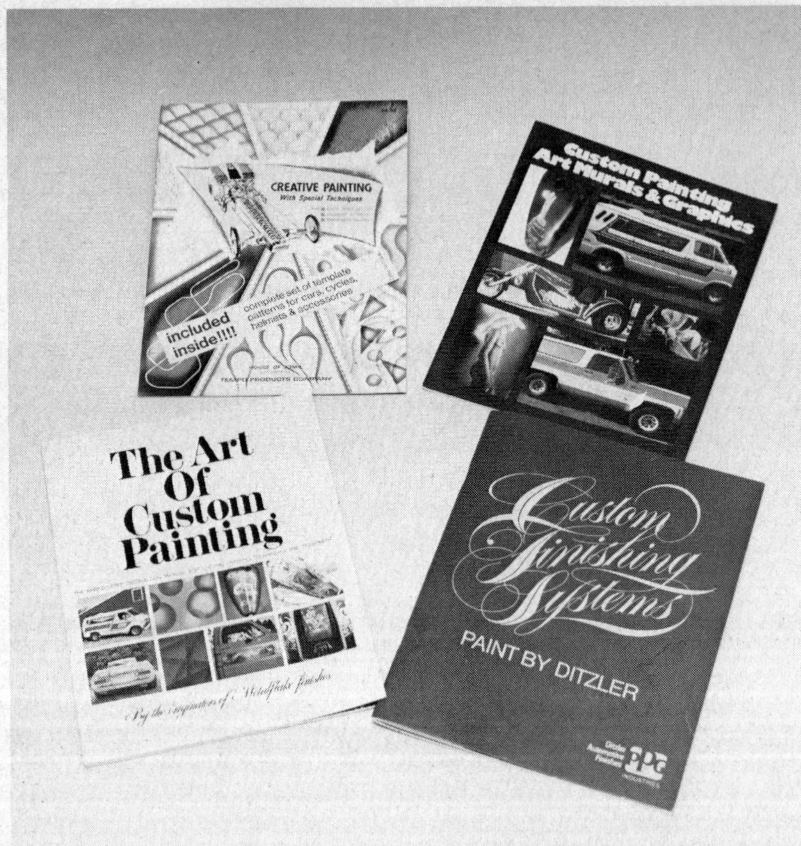

Several paint companies publish very informative booklets to aid the customizer. Clockwise from lower right, the companies are Ditzler, Metalflake, Tempo, and Custom Productions Inc.

WASH PRIMER–
Used to protect bare steel surfaces against corrosion. It should be covered with enamel primer or lacquer primer-surfacer, depending upon which material is chosen for the topcoat.

PRIMER–
A non-sanding coating designed to provide a good base for the topcoat. The pigment content is low, therefore sanding scratches on the surface will show through easily.

PRIMER-SURFACER–
A heavily pigmented primer used to fill in sanding scratches and minor surface imperfections. Wet sand with #400 paper before applying topcoat.

SEALER–
A non-porous coating to be used over old paint if a dissimilar type of paint is to be sprayed on top of it. Can also be used to prevent bleed-through of a previous color in a repaint. Normal primers, being porous, are not adequate substitutes for sealers.

METAL-PREP–
Diluted with water, this liquid is used to remove rust and corrosion from bare steel before priming. It also etches the metal, providing a tooth for the primer.

ALUMA-PREP–
Similar to Metal-Prep, but used for bare aluminum. Specifically formulated to remove aluminum oxide which forms rapidly on unprotected aluminum. Must be followed by prime coat of zinc chromate.

ZINC CHROMATE–
A special primer used to protect aluminum from corroding or oxidizing. It is compatible with all primers and primer-surfacers, although priming is not necessary before the application of enamel topcoats. Again, because of their porosity, normal primers are not an adequate substitute for this product.

ENAMEL–
The most common type of automotive finish paint, it is traditionally composed of a pigment, a resin or binder, and an oil-based vehicle. Enamels dry slowly by a combination of evaporation, oxidation and chemical reaction. Noted for having excellent hiding and filling properties. Glossy when dry, requiring no further finishing work. Modern synthetic enamel with an acrylic resin and no oil is called ACRYLIC ENAMEL. If a catalyst is added to speed up drying, the product is known as TWO-PART ENAMEL.

LACQUER THINNER–
A solvent used to lower the viscosity of lacquer and lacquer primer so it can be applied with a spray gun. There are many types to choose from and correct selection of the proper thinner can yield an improved finish. Extra high gloss thinner, spot repair thinner, non-penetrating thinner and general purpose thinner are just a few of the many types offered.

ENAMEL REDUCER–
A solvent used to lower the viscosity of enamel and enamel undercoats so spraying equipment can be used. As with thinners for lacquer, many types of enamel reducers are available.

RETARDER–
An additive that slows evaporation and prevents blushing of lacquer topcoats when weather is extremely warm or humid. For enamel, reducer/retarder increases flow in hot weather conditions.

LACQUER–
A paint system invented by DuPont in 1922. It consists of pigments, resins and solvents. It dries rapidly by evaporation of the highly volatile solvents. Drying hard and brittle, it requires a rubout and polishing to achieve that brilliant depth and luster for which it is famous. Early types which left a nitrocellulose film were called NITROCELLULOSE LACQUER; modern versions with acrylic resins are ACRYLIC LACQUERS.

POLYURETHANE ENAMEL–
A two-part synthetic paint which hardens by chemical reaction between the two base components, which are mixed just before application. Polyurethanes have exceptional resistance to weathering, abrasion and chipping. A recent variation with a base acrylic resin is ACRYLIC URETHANE.

ELASTOMERIC ENAMEL–
A two-part (catalyzed) enamel containing ingredients that give the enamel great flexibility. Commonly used on flexible bumpers, fender flares, etc., where collision will compress or flex part. Paint will return to normal appearance when base part recovers its original shape.

EPOXY–
Another two-part paint in which a catalyst is added to a pigmented epoxy resin just before application. Very hard, high gloss, but limited color range. Once favored for chassis parts, epoxies are losing out to polyurethanes.

METALLIC–
The presence of metallic powder in the tinted base paint. Custom metallic shades are used as an underbase for candy colors and pearl finishes. The powders are available in grades from fine to extra coarse in gold, copper and silver. Many colors are available ready-mixed from paint dealers, but custom blending with powders, toners and clear blenders is still popular. Photo 7 on page 169 reveals the beauty of metallic finishes in color.

CANDY–
A translucent finish obtained by mixing colored toner into clear and adding modifier. Candyapple red, for example, is blended from 1 qt. DuPont maroon toner, 1 qt. Acme 83H modifier, 2qts. Acme acrylic 607 clear and 3 gals. quality thinner. Since the paint is translucent, each succeeding coat deepens the color when sprayed over a metallic silver base, the finish obtained resembles a candied apple. A color example of candyapple is to be seen in photo 1, page 170.

PEARL–
Today's pearl pigment is made by treating tiny flakes of mica with lead carbonate, or better yet, an acid solution to produce a finish with a soft, iridescent sheen when sprayed over a white base-coat. Pearl can be mixed with candy colors for unusual two-tone effects. Metalflake, Inc., offers this in a pre-packaged product called Flip-Flop; as does Aero-Lac with their Pearlescence Colors. Tinted pearl in slow-drying mineral spirits base are likewise packaged by both companies. Otto Rhodes' truck (photo 6 on page 171) proves that one color picture is worth 1000 words when it comes to describing a straight pearl finish over a white base.

METALFLAKE–
Tiny rectangular bits of colored Mylar added to clear, with or without toner, produce this highly reflective finish. (Refer to page 171, photo 7.) The color, size and quantity of flake can be varied to create any desired degree of reflectivity. Constant agitation is required to keep the flake in suspension while spraying, and then many coats of clear are needed to seal the flake and smooth the surface. Although Mylar flake is offered by many companies (for example, Aero Flake from SEM Products, Inc.), Metalflake, Inc., produces the original Metalflake and its companions Spindrift and Mirra.

CERAMITATION–
Trade name for a super-hard two-part enamel from Tivian Laboratories, the possibilities of this ceramic-hard enamel (can be stoned and polished when fully cured) have yet to be exploited by auto painters and customizers. The range extends through 56 standard colors including metallics and fluorescents. It can be sprayed or brushed on metal, plastic, fiberglass, wood and other materials.

Airbrush Tips and Techniques

The special equipment needed for spraying fine lines and color gradations.

by Al Hall

The airbrush is a commercial artist's tool, and as such sees wide use in photo retouching, advertising art and technical rendering. Industry has borrowed the artist's airbrush for painting toys, ceramics, and countless other products on an assembly-line basis. Custom car and van painters have also taken a great liking to the airbrush because of its ability to deliver a variable pattern of lacquer, or enamel: the automotive painter's favorite mediums.

Spray guns and airbrushes are similar, but by no means identical except for size. Although both use air to atomize and deliver paint to the surface, each controls this process in a different manner. An airbrush gives the painter the ability to vary the pattern as well as the volume of paint. There are two basic styles of airbrushes: internal mix and external mix—just as with spray guns. There are also airbrushes with needles and without needles. The latter are invariably single action and the former are double action.

A single action airbrush features a top-mounted button that controls air volume and a separate adjustment (usually a knurled, screw-type fitting below and/or in front of the air outlet) for paint volume. This means using a single-action airbrush is a two-handed operation. The right hand holds the brush, with the right index finger controlling air volume as the fingers of the left hand manipulate the threaded sleeve that varies the paint volume and thereby the spray pattern. A needle-less, single-action airbrush (such as those offered by Badger, K.J. Miller, and Metalflake) can spray coarse, heavy pigmented paints—even Metalflake—that the fixed-needle, single action airbrushes can't handle. And pattern width can be varied (though normally not while spraying) from a 4-in. width down to ½-in. wide or a bit less. These needle-less, single-action airbrushes are the least expensive to purchase (the Miller Model 2011 outfit with powerful compressor, airbrush and accessories costs less than many double-action airbrushes alone), and easiest to clean. Their disadvantage is in being unable to spray a small pattern and in the difficulty encountered going from broad to narrow or vice-versa.

A needle-controlled, single-action airbrush does away with the difficulty of being able to spray a small pattern and minimizes clumsiness of quickly going to broad coverage from narrow or vice-versa. It is not quite as easy to clean as the needle-less model, but counters this disadvantage with its ability to utilize paint cups as well as various size bottles. This feature is quite handy when doing murals since the colors desired for the mural can be placed in separate bottles according to volume of paint

1

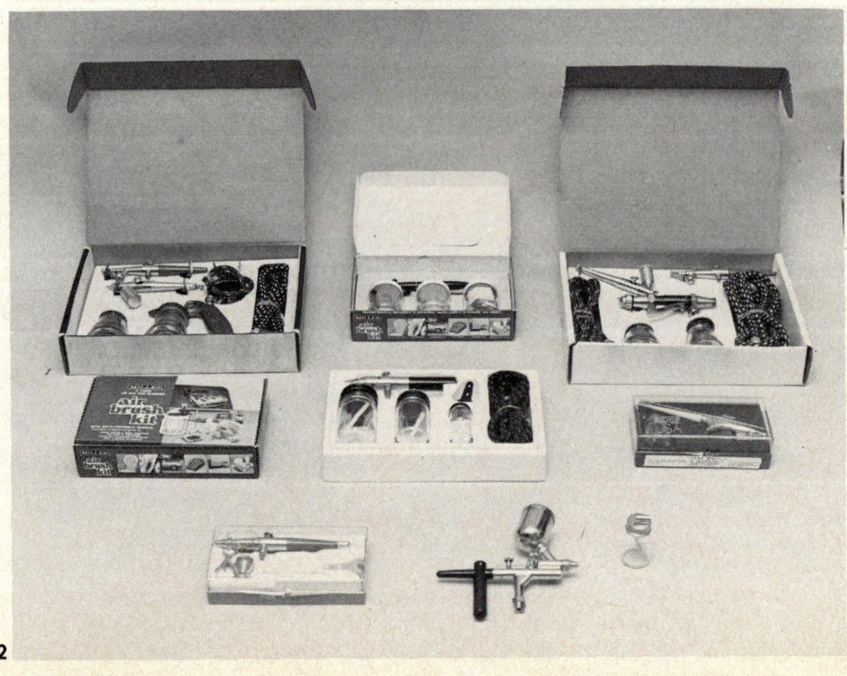
2

needed and instantly plugged into the airbrush. If a little touch-up on the mural is required and the amount of paint in the bottle isn't enough to allow pickup by the siphon tube, it can be poured into the cup and the job finished without having to mix up another batch of paint. A needle-controlled, single-action airbrush can quickly go from a 2½-in. broad swath of paint to a 1/16-in. wide line.

A sliding-needle-controlled, double action airbrush is considered by most artists to be the epitome. The top (or side) mounted button controls both air and paint volume. Pushing down on the button increases air flow, while pulling back increases paint flow. Thus, simultaneously pushing down and pulling back increases the spray volume and broadens the spray pattern. The pattern change is effected because pulling back the button moves a long, tapered needle in the barrel of the airbrush away from the orifice. These double-action airbrushes are internal mixing, hence, cleaning them means complete disassembly. And that long, tapered needle means these airbrushes are more delicate than their kin. It also means coarsely pigmented paints will cause considerable problems with clogging. Only finely pigmented enamels, lacquers, etc., are suitable for use in a double-action airbrush.

Query a commercial artist about the best airbrush and the response will probably be: "Oh, a Wold Model A-1 or BBF, a Thayer & Chandler Model A, a Paasche Model AB, or Badger 100XF, or perhaps a German-made Efbe!" Unfortunate-

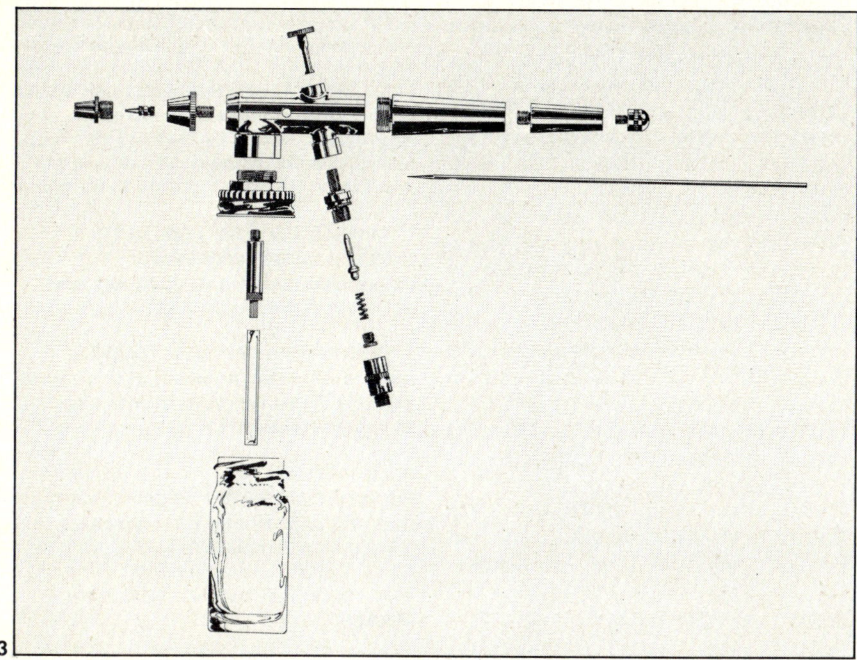

3

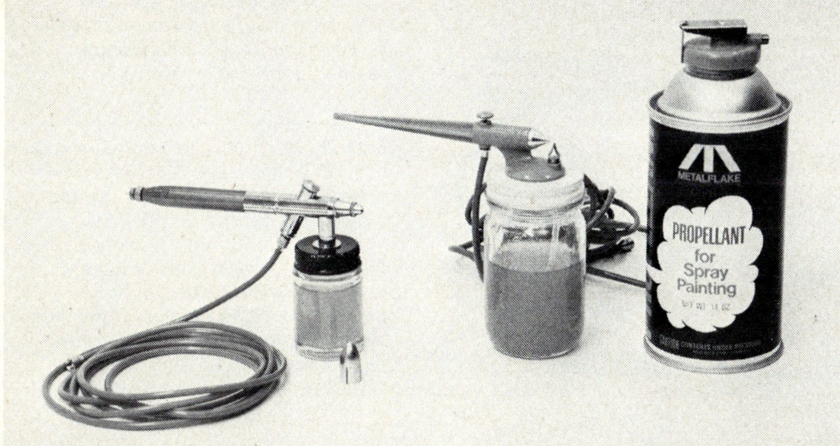

4

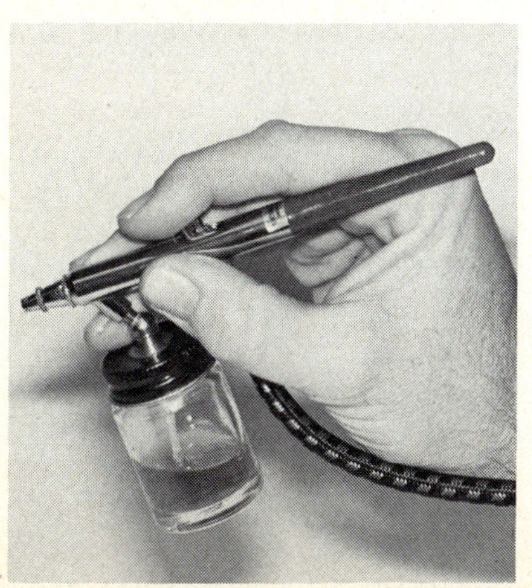

5

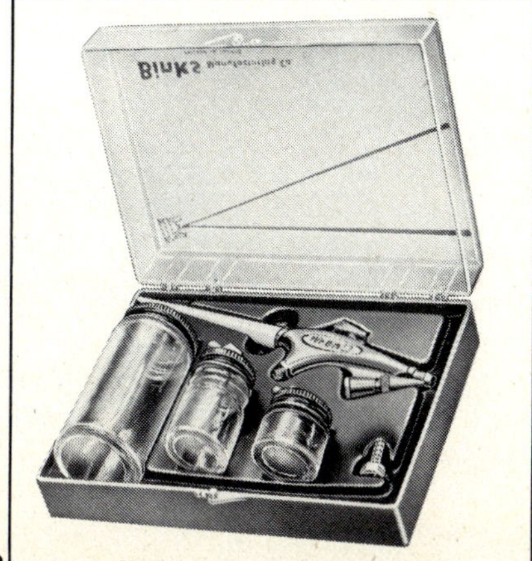

6

1. Tom Kelly of The Crazy Painters adds the finishing touch to a diffraction tape logo on the back of a Corvette with a . . . you guessed it, Airbrush! One of the most versatile tools in the painter's bag of tricks. This is a Paasche Model VL.

2. Just a few examples of the many brands and models of airbrushes suitable for custom automotive use. These are all available by mail from Seelig's Custom Finishes. They are (clockwise from top center): K.J. Miller single-action needle-less; Paasche Stripe N' Spray Touch Up Kit with Model H #3L airbrush, AUTF-2 thumb-action spray gun and flow pencil; the Binks Wren airbrush, the German-made Sata Universal #400, Paasche's Model VL; K.J. Miller AB-200; Paasche Auto & Hobby Paint Kit, and the middle unit is a Thayer & Chandler Model E single-action airbrush kit.

3. Exploded view of single-action Thayer & Chandler Model E. Not all of these components need to be disassembled for cleaning. Note the long, thin needle at right. It's the most delicate airbrush component.

4. Metalflake Incorporated sells the needle-less MiniGun (right). Yes, with optional tip, it can spray Metalflake! The SuperGun (left) is a sophisticated needle-type single-action airbrush.

5. Metalflake's finest airbrush is the superb double-action ProGun. This is proper way of holding an airbrush. Note that hose is wrapped over forearm so it won't drag in the paint.

6. The single-action, external mix Wren model airbrush by Binks has flattened sides, squarish button. It's unusually comfortable to hold and doesn't tire the hand during prolonged use.

ly, none of these is really suitable for custom car and van painting. They are all double-action artist's airbrushes, and while capable of painting a line as fine as that made by the sharpest pencil, they are not suitable for shooting lacquers or enamels.

AIRBRUSH CHOICES

The best model Wold double-action airbrush for automotive use would be the Master M or Model Master. Thayer & Chandler's Model C, Paasche's Model VL or V, Badger's 100, The Metalflake ProGun and the sensational Sata Universal #400, are likewise excellent for applying lacquers and enamels.

When it comes to single-action, needle-controlled airbrushes, the custom car and van painter should choose from among the Wold Model U, Thayer & Chandler Model E or G, Paasche Model H, Binks Wren model, Badger Model 200 and Metalflake SuperGun. Pricing in this category of airbrushes runs from just over $25 to just under $60, depending upon the brand, model and accessories selected.

Although strictly speaking they are not airbrushes because they lack needle control, the needle-less guns should not be lightly dismissed. Their ability to handle heavy materials (even Metalflake) and easy cleaning make them of great value where ultra-wide coverage is unwieldly and ultra fine spray patterns aren't needed. Foremost among the brands in this category are K.J. Miller's AB100 (bleeder type) and AB200 (non-bleeder), Metalflake's Mini-Gun and Badger's Model 250. Since most murals are done with the aid of masks or stencils any of these guns can handle the job easily—in addition to pattern spraying.

Take a poll of professional muralists and custom painters and it's likely virtually all the airbrushes previously mentioned will be named. Tom Kelly (of The Crazy Painters fame) prefers the Paasche VL. Steve Young (likewise one of the Crazy Painters) opts for the Paasche Model H single-action airbrush. Carl Caiti (The Psychedelic Psycho from Florida) alternates between Metalflake's ProGun and their SuperGun. The famous Gene Winfield seldom does small detail painting, so the Binks is more than adequate for his needs.

AIR SUPPLY

Although many of the aforementioned airbrushes are sold in kits

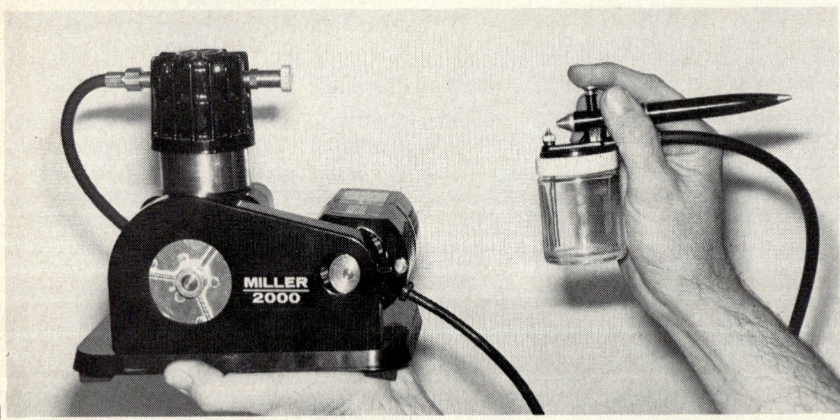

1

2

3

1. K.J. Miller's needle-less airbrush easily sprays lacquers and enamels—with larger paint nozzle it'll even spray Metalflake. The palm-sized compressor is a fooler. It can put out up to 50 psi.

2. Thayer & Chandler's Model E is excellent quality single-action, internal-mix airbrush for lacquers and enamels: A professional tool.

3. Double-action Paasche Model VL is a favorite of many professional van mural painters. Three separate needle sizes give great versatility.

4. The Stripe N' Spray Touch-Up Kit from Paasche includes thumb-action spray gun with 3-oz. siphon jar, Paasche's unique Flow Pencil with 1/32-in. striping tip and Model H single-action airbrush, 3-oz. jar and hoses plus instruction booklet.

5. Using frisket paper to make stencils or employing reusable masks for airbrushing murals means painter can use less-expensive airbrush since fine-line control isn't needed.

6. One manufacturer's airbrush generally won't fit onto another maker's hose. All of these are different enough to prevent interchange. They are (right to left): clear plastic air hose (not recommended for high pressures used with auto paints); K.J. Miller hose, braided Thayer & Chandler hose; Paasche's braided hose; braided hose for Binks Wren airbrush, and braided hose for Metalflake's ProGun.

with aerosol cans, these are generally to be avoided unless it's to be a one-time job. The commercial artist generally prefers a bottled air supply. This is merely a carbonic air tank filled with CO_2 to avoid contaminating the paint with moisture. If a portable supply is necessary then this is a good choice. Ditto for one of the miniature compressors (for example, the K.J. Miller Model 2005 with adjustable relief valve and ⅛th horsepower motor). But by far and away the most popular and workable air supply is the shop compressor with suitable moisture trap. The simple fact of the matter is that an airbrush can utilize the same air as a standard spray gun. This should be good news to those who suspected these fussy-looking little contrivances of requiring something exotic—and expensive. It isn't a bad idea to add a small in-line moisture trap close to the airbrush (Paasche offers a good one) if there's any great distance between the airbrush and the compressor.

HOSES & FITTINGS

There are two basic kinds of hoses used with airbrushes. The first is a clear plastic tube that comes with most of the low-cost units. This kind of hose is fine for hobby and craft work with painting mediums such as dye or ink, etc., since the pressures are very low. For lacquers, enamels and typical automotive finishes where pressures are high (20 or 25 psi all the way to 60 psi), the cotton-braided rubber hose is a must. The better and/or more expensive airbrushes provide these as standard. Be informed, however, that while the compressor end of the hose is equipped with the standard ¼-in. fitting, connectors for the airbrush ends are not standard. That goes for paint cups and bottles, too. So unless fancy silver soldering is a hobby, make sure the airbrush selected for use can be mated with bottles from the manufacturer in sizes from ½- or ¾-oz. to 2-, 3- or 4-ozs, and cups of ¼- and ½-oz. capacities. It would be nice if these cups had optional caps so spillage would be no problem, but this is rare. Only the German makers, Sata and Efbe, provide covers for their airbrush cups—although Paasche incorporates one on their flow pencil. Airbrush accessories include everything from handy tote caddys to organize brush, bottles, cups, hose and fittings, to in-line needle valves to control unregulated air supplies, and fittings to permit using two or three (or more) airbrushes on the same compressor outlet.

AIRBRUSH PAINTING

Proficiency with an airbrush is mostly a matter of practice, practice and more practice. That is if the user is posessed with a modicum of artistic talent. A lack of the latter can be overcome to a great extent by following the instructions in one of the several good books on the subject. These instructions are devoted to learning how to achieve proper shading on geometric shapes, airbrush straight

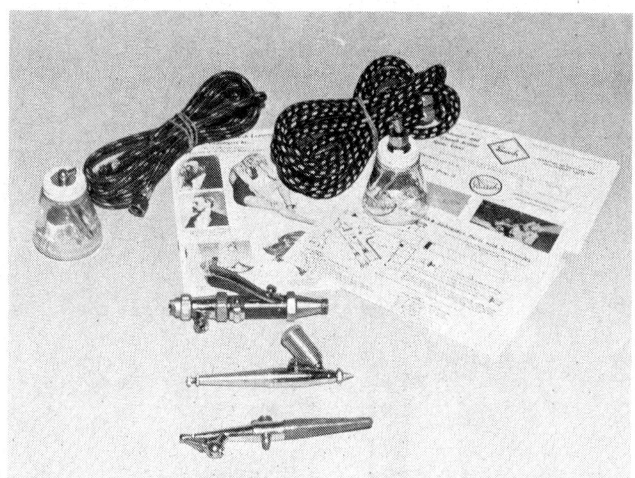

4

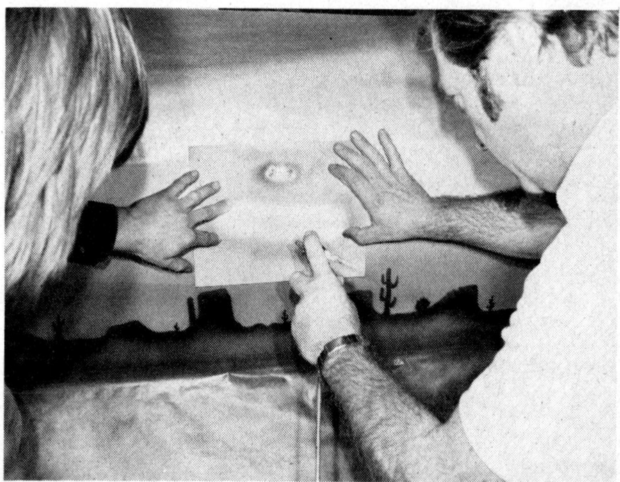

5

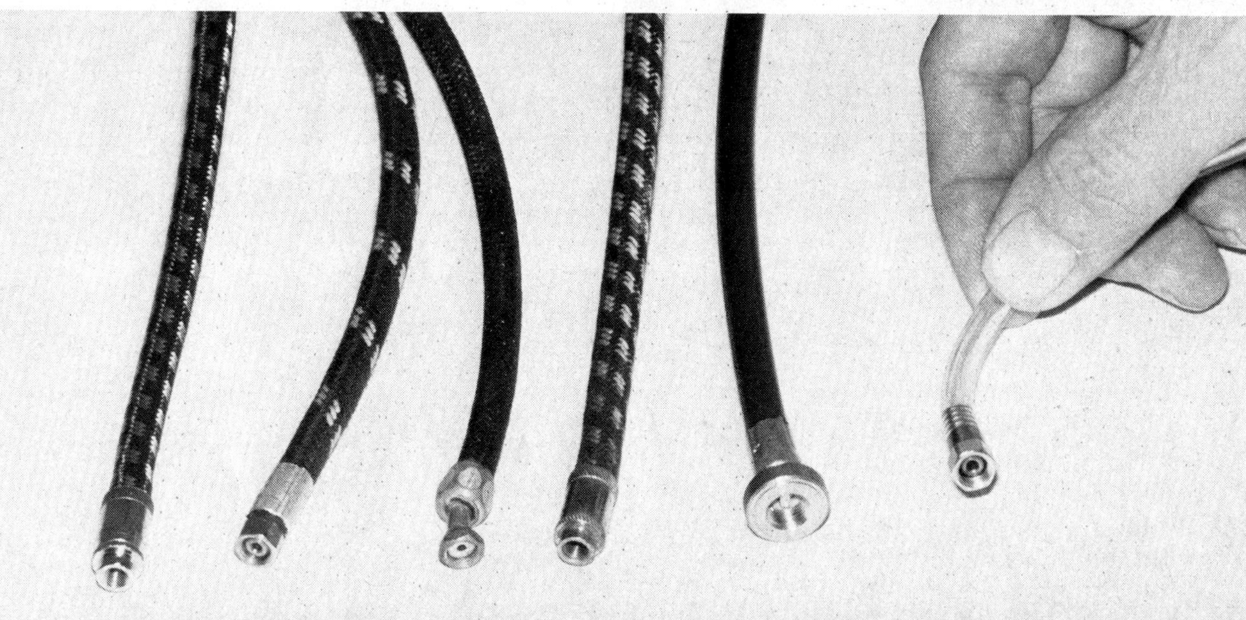

6

lines of varying thichnesses and special techniques such as stippling and spattering. A book highly recommended, in addition to those illustrated in this article, is "Complete Airbrush Techniques" by Sol Dember (published by Howard W. Sams & Co.) which includes a very good chapter on color techniques.

It certainly isn't much fun to practice putting evenly spaced dots on a sheet of paper with an airbrush, then connecting these dots with the finest possible straight lines that do not vary in thickness and finally shading specific areas with repeated passes to gradually build up darker tones. The result of this practice becomes mastery over the airbrush and means the airbrush painter can walk over to the side of a vehicle, push the button and know how much paint will be applied to the surface—and exactly where it will be applied. The secret of creative custom painting is in creating a three-dimensional image on a two-dimensional surface. This requires a mastery of the techniques of toning and shading, and these effects are so easily and quickly achieved with an airbrush. Just consider the cost of a mural on the side of a van if it had to be

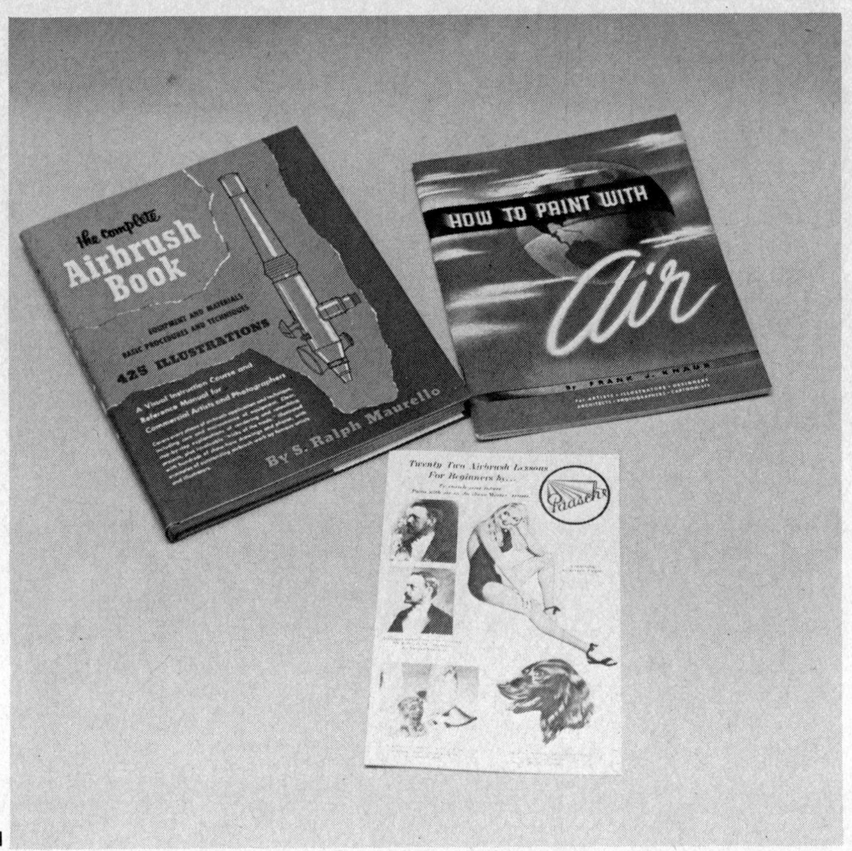

1

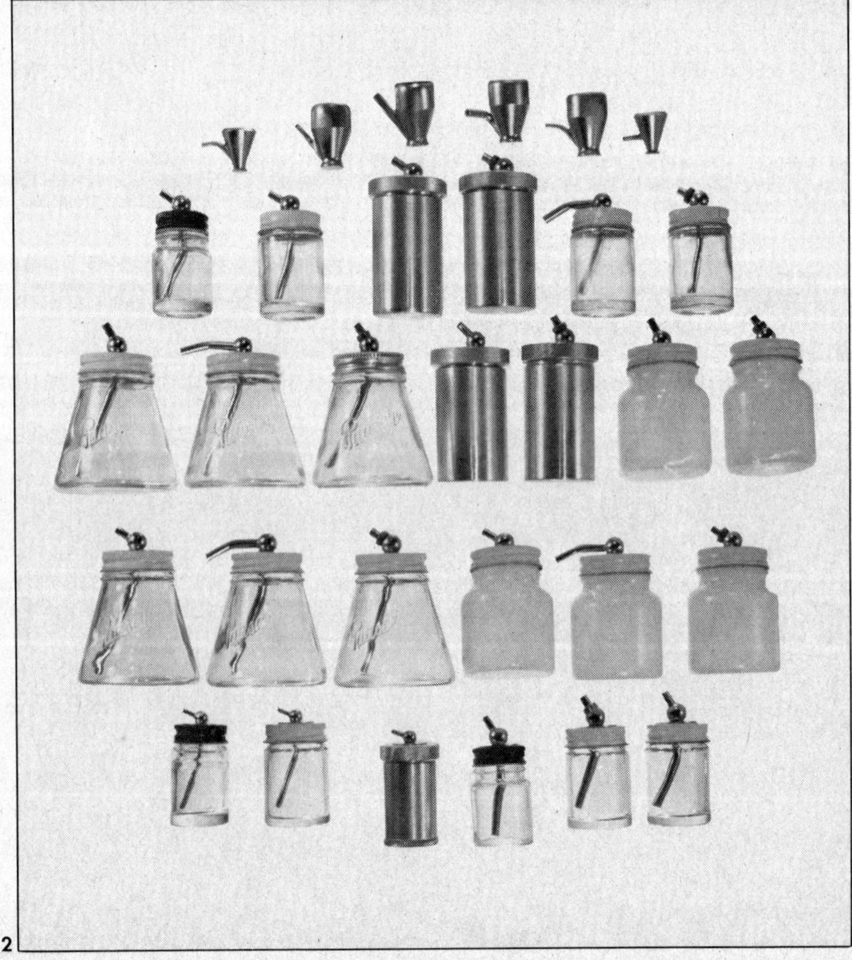

2

1. Good advice for the novice can be found in these books on airbrushing. They are: The Complete Airbrush Book by S. Ralph Maurello, How to Paint With Air by Frank J. Knaus and Paasche's useful Airbrush Lessons for Beginners.

2. Put your paint in here. Choose from small, medium and large cups or varying size bottles in glass, plastic or even metal from Paasche.

3. Other airbrush accessories include inline moisture filters, various braided hoses and connectors, adaptors and multiple air outlets. These are just a sample from Paasche.

4. Binks' Wren model can be adjusted to spray very fine pattern. Large color bottle is handy for big jobs.

5. This kit by Thayer & Chandler includes Model E single-action airbrush, three color bottles and quality braided air hose. Oh yes, there's also a clip to hold brush.

6. Paasche makes some very specialized units such as this VL airbrush with remote one quart color bottle and air regulator/moisture trap for use with carbonic air bottle.

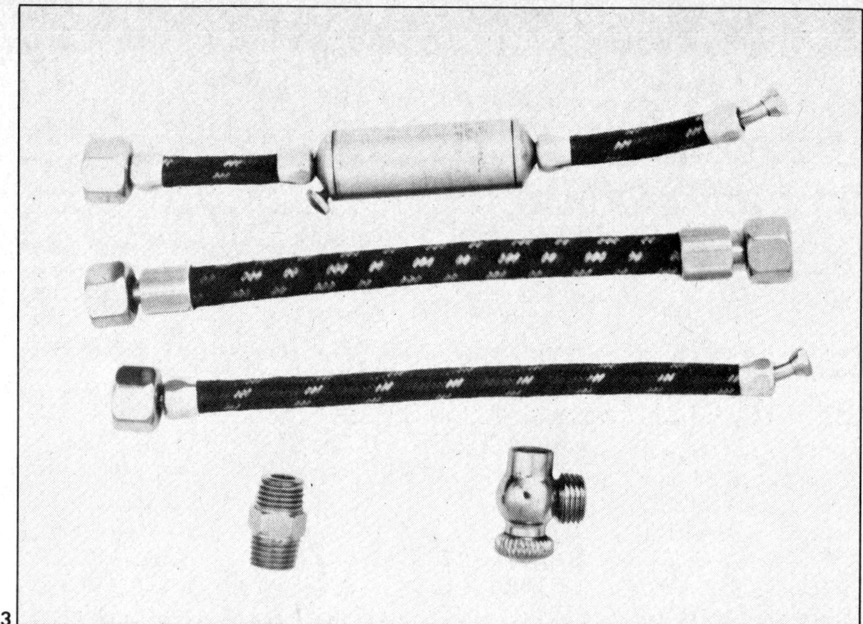

3

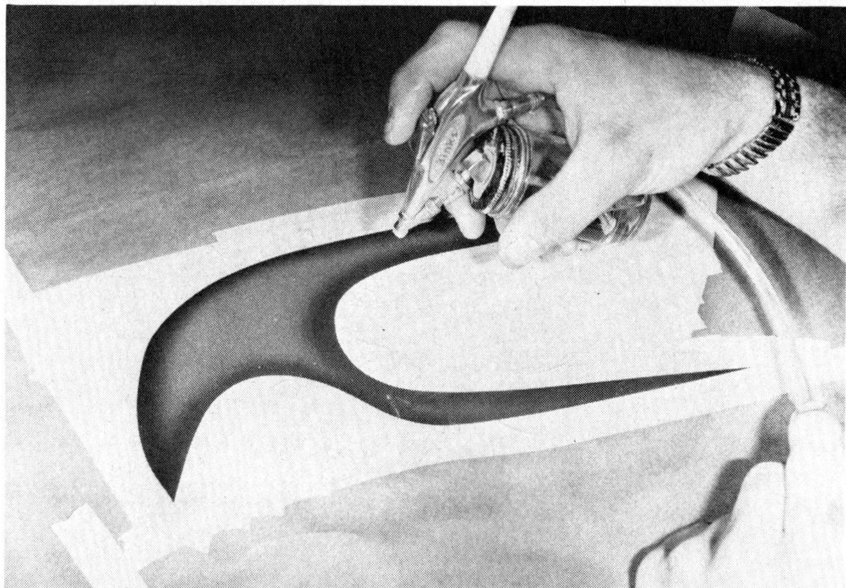

4

done with camel-hair brush and palette. Could you afford it? Could your customer afford it?

One of the greatest difficulties a novice airbrush user encounters is in gauging the proper spraying pressure. This might vary from airbrush to airbrush and is greatly dependent upon the type and even the brand of paint used. As a guide, dilute the paint until a drop placed on a smooth, vertical surface runs freely, but leaves behind it a trail of color the same width and fairly heavy. Set the regulator on the compressor at 35 or 40 psi if the hose is short and higher if there's a lot of hose between the compressor and the airbrush. Make a trial spray on a sheet of cardboard. If the result is spattering, add a bit more thinner to the paint, stir well and try again. If it still spatters, increase the air pressure and try again. When satisfactory coverage is obtained, make a written note of the ratio of paint/thinner and air pressure setting for future reference.

If, on the other hand, the spray pattern is blotchy and runny, add more paint. If the texture is grainy, it's most likely because the paint is too thick or not thoroughly mixed. If, instead of a smooth swath of color, the pattern resembles a centipede, the air pressure is too high or the airbrush is being held too close. When switching from one color to another, always spray a short burst of thinner through the airbrush. This is easily done if a separate bottle of thinner is kept handy.

MAINTENANCE

Just swishing a little thinner

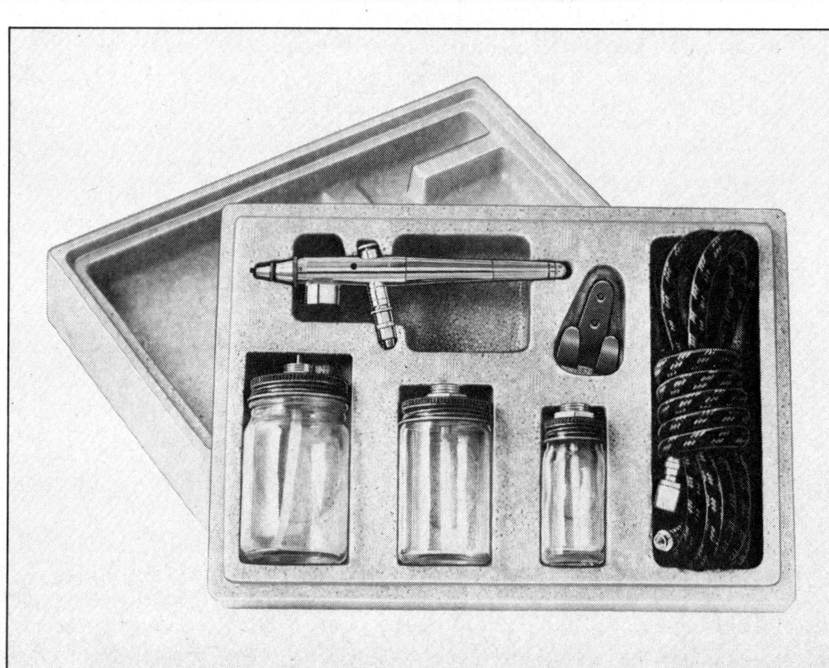

5

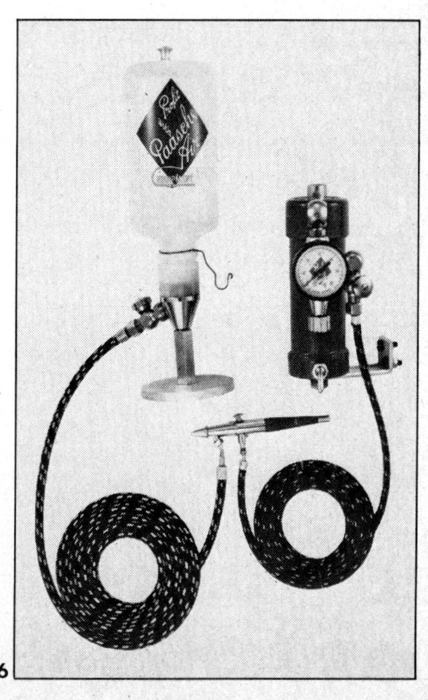

6

through the airbrush after the job is finished won't get it though. Since even minute particles of dried paint can affect the performance of an airbrush, it should be disassembled and thoroughly cleaned after each job. On a needle-less single-action unit, this involves nothing more than washing four or five parts out with thinner and coating the threads on the paint nozzle with bee's wax after everything is thoroughly dry.

A needle-controlled, single-action airbrush (external mix) can be separated into slightly less than 10 components in need of cleaning, but great care must be taken not to damage the delicate point of the needle. A thorough cleaning requires lint-free rags, a bristol brush, round toothpicks, a few cotton swabs and some soft pipe cleaners. The color cup or bottle is washed out with thinner first. A rag dampened in thinner can be used to clean around the threads and the air bleed hole. Pipe cleaners are used to cleanse the paint outlet pipe on either the cup or bottle lid. Some cups have their own needle and seat, in which case the needle is removed and both cleaned with the bristol (round, moderately coarse bristle) brush. On the airbrush itself, the color adjusting needle and seat are loosened (usually by threaded end or Allen-head set screw) and disengaged from either side of the airbrush shell. A small-bladed screwdriver may be needed to remove the packing nut and packing washer. If the needle and tip orifice don't respond to cleaning with the bristol brush, use nothing harder than a wooden toothpick to rid the orifice of paint. Soaking in a jar of thinner will usually soften paint residue and allow it to be flushed away. Usually the only other item requiring removal for cleaning is the airbrush tip, or air cap as it's called. Lubricate the air cap threads and the threads on the color adjusting parts and packing nut with the tiniest bit of Vaseline or bee's wax to prevent paint

1

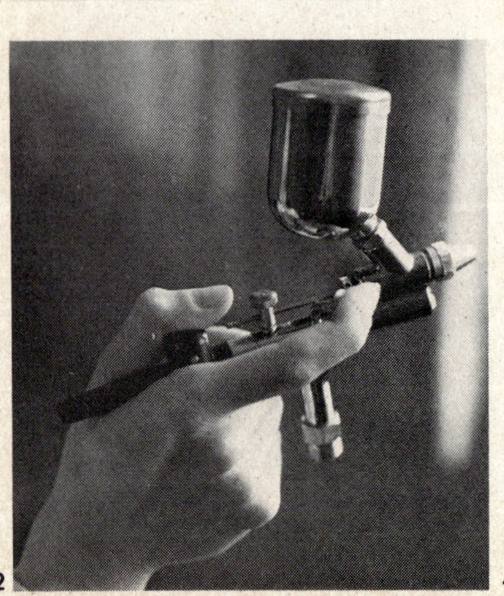

2

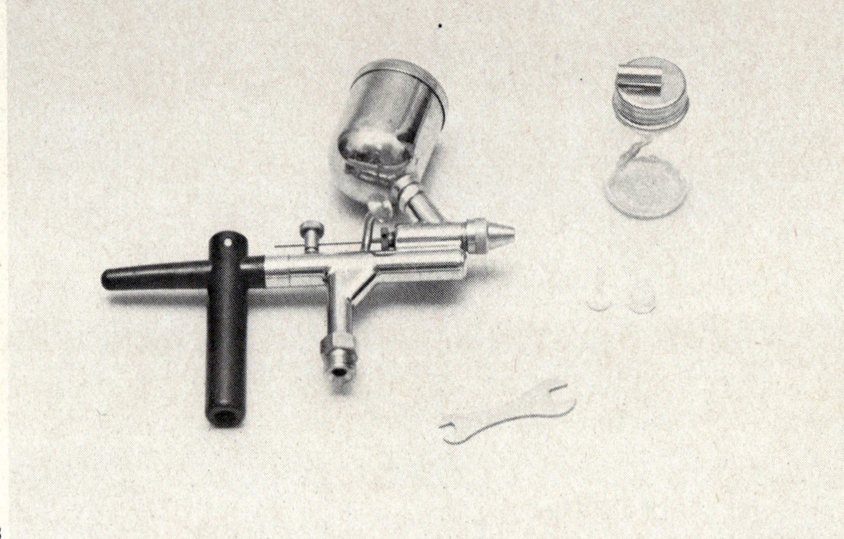

3

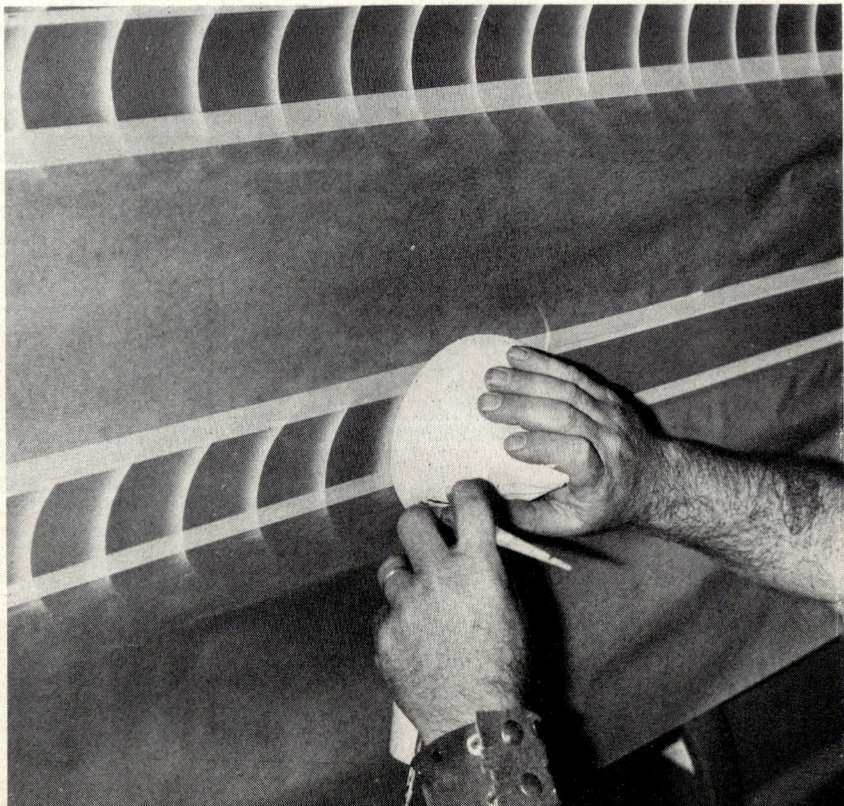

4

buildup from freezing these components.

The internal-mix version of the aforementioned has, at the most, slightly over 10 parts that would be involved in a cleanup program. The most sensitive of these is the long, thin needle. If the ultra-fine point on this needle gets bent, try straightening it by rolling it on a sheet of glass (or equally smooth, hard surface) and stroking it from back to tip with the back of a fingernail. If that won't straighten it, the only other choice is to purchase a new needle.

Cleanup on a double-action airbrush with its assortment of 20-or-so teeny parts is only dreaded by most airbrush painters because of the chance of losing or damaging a part. Disassembly and reassembly cannot be rushed and should be done on a clean section of the workbench. A lip along the edge will help to keep parts from rolling away and getting lost. Before disassembling the airbrush, thoroughly clean the color cup or bottle and work paint out of attaching hole in airbrush with bristol brush and thinner. Then pour some clean thinner in the cup or bottle, attach to airbrush and back-flush by placing a finger over the airbrush nozzle and moving the button all the way down and back. This flushes any remaining paint backward through the airbrush into the color bottle or cup. Play it safe, reduce air pressure at the compressor and hold the thing under the workbench if back-flushing through the color cup. Repeat procedure if paint color shows up in cup or bottle until no paint shows. To field-strip the little jewel, follow the instructions packed with it at time of purchase.

ALSO OF INTEREST

A close cousin to the airbrush is finding great favor among custom enthusiasts. It's manufactured by Paasche and called an Air Eraser. It looks and handles just like an air brush, but it uses controlled air and a fine abrasive compound to help commercial artist's eliminate mistakes in renderings, confusing backgrounds, etc. The automotive shop can increase air pressure and use Paasche's Fast Cutting Compound to remove small patches of surface rust, certain types of paint blemishes and even etch designs in window glass. Since it's basically and airbrush with a carballoy tip, it can etch fine-line designs with ease, then open up for wider background work.

Paasche's other cousin to the airbrush is called a Flow Pencil. It's a gravity-fed pinstriping machine that's as easy to handle as a pencil. The color (enamel or lacquer) is mixed to free-flowing consistency and poured into the cup. The lid is secured and a puff of air into the cup starts the flow—no air compressor needed. The Flow Pencil is touched to the surface and the flow control lever pulled back. The width of the stripe can be varied by selection of one of four different tips (1/64th-, 1/32nd, 3/64ths- or 1/16th-in.) and adjustment of the color adjusting screw. An adjustable, revolving swivel guide aids in maintaining proper spacing.

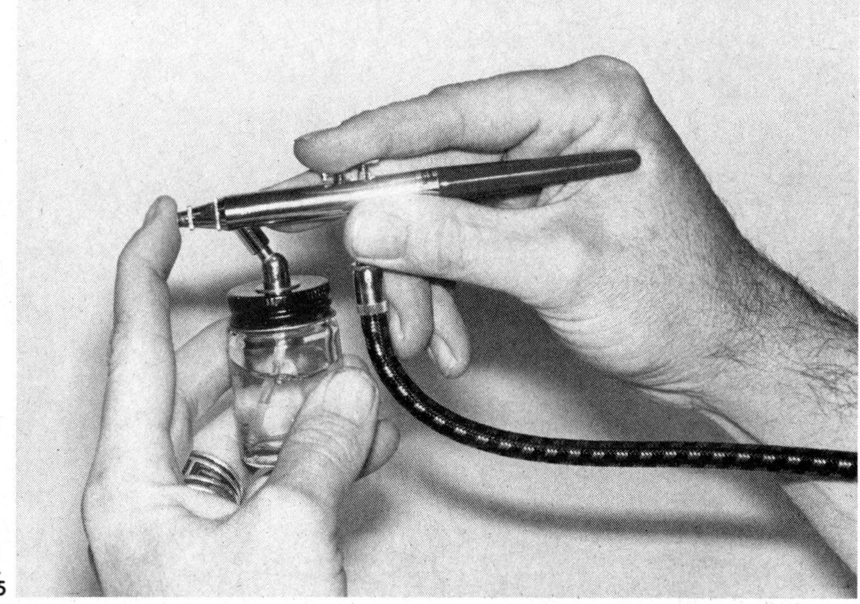

5

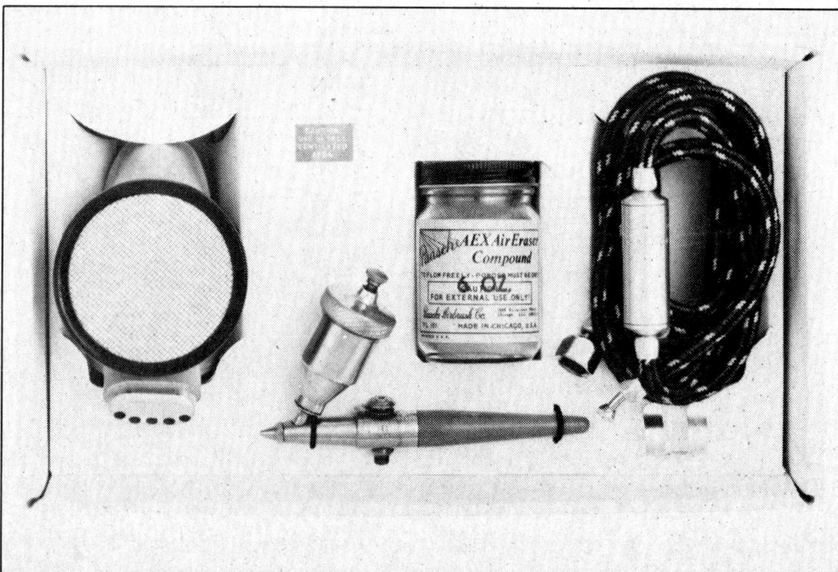

6

1. Thayer & Chandler's new Model C is an ultra-professional double-action airbrush engineered especially for use with automotive paints.

2. This is German engineering turned loose on airbrush design. The Sata Decor-Z Universal #400 features adjustable sliding handle, positively huge color cup with lid, and preset spray pattern setup.

3. The Sata can also utilize a color bottle. Unusual slide-action lever (button) and adjustable sliding handle can be seen here.

4. An airbrush (Binks Wren), a paint strainer and presto; a trick pattern takes only minutes.

5. Quick cleaning job on double-action (Metalflake ProGun) airbrush is performed by filling color cup or bottle with thinner and holding finger over tip while pulling button down and back to backflush airbrush. Repeat with fresh thinner until no more color appears in jar.

6. It's made by Paasche and it looks like an airbrush, but it's a transistorized sandblaster known as the Air Eraser. Etching glass was never so easy, quick or controlled.

Tapes For Custom Painting

Whether you want to secure, separate or decorate, there's an adhesive-backed product to do the job

by AL HALL PHOTOS BY PAUL DEXLER AND ERIC RICKMAN

There are many types of tape and probably a hundred times as many applications, but uses can be grouped into three basic categories: 1) binding, 2) protection, and 3) decorating. Since package wrapping and quickie body repairs aren't under discussion, that dispenses with category one.

The second category makes sense if protection is thought of in terms of protecting chrome, windshields, tires, etc., against overspray. Protection is also needed when applying paint designs so that the borders of the design are defined and adjacent areas protected from overspray. There are probably many other reasons for labeling it protection, but these will suffice for now.

MASKING TAPE

Where would a painter be without masking tape? Even if it were possible to completely strip the vehicle of all items not being painted, masking tape would still be necessary to provide separation between colors and outline areas for application of paint designs. However, it is not possible to completely strip a vehicle for repainting, so masking tape remains the painter's friend. Available in widths from ⅛-in. up to 4-ins. and beyond (for special applications), the most useful sizes for custom painting are ⅛-in., ¼-in., ¾-in., and 2-ins. This selection will handle most custom painting tasks. Of course masking paper is almost as indispensable as the masking tape. For best results, the tape should overlap the paper edge by 50%. A masking paper taping dispenser applies tape with just the precise amount of overlap automatically as the paper is pulled off the roll. These machines are costly, but worth it.

The most common mistake made by the amateur custom painter is in buying bargain-table rolls of masking tape or spending more, but getting household quality masking tape. Don't waste money on these products. 3M's Scotch Brand Pressure Sensitive Masking Tape for automotive use is expensive, but in comparison to the other materials required represents a fraction of the cost. A good piece of advice is to always use the best quality materials. Then if the finished job isn't first rate; the fault can only be with the painter not the products he used.

High on the list of common masking mistakes is not getting the surface clean right to the base upon which the paint—and masking tape—will be applied. All wax and polish must be completely removed. This is especially important if the old top coat has been treated with a silicone. Masking tape (and that new coat of paint) will not adhere to a silicone treated surface. Since it goes on first and comes off last, mistakes in masking are especially difficult to detect during painting—only after the tape is removed will the color that crept beneath it be noticed. This is caused by another of the most common masking mistakes: Not making certain the edge of the tape is burnished down. Seating ⅛-in. and ¼-in. tape is best accomplished with a small roller. Roll it over the tape several times, starting with light pressure and gradual-

ly increasing it. If excess pressure is used on the first pass, the tape may squirm or shift.

DESIGNS IN TAPE

Panel outlines, whether for flames, borders, etc., are easy to lay out with ¼-in.-wide masking tape. With this width it's easy to run curves down to 3-in.-dia. without the inside edges puckering. When pulling a straight line down the side of a vehicle, the best technique to follow is to firmly seat the first few inches, then holding the tape away from the vehicle a few inches, spool it out the full length of the line. Apply slight pressure to stretch the tape a bit, and lay it straight down. If the result isn't a dead straight line, lift, stretch and lay straight down once again. If satisfied, use the roller to seat it. The ¼-in. wide masking tape comes in quite handy for taping body side lettering (if removal is impossible) and in conjunction with ⅛-in. tape for separation stripes and endless line designs. (This latter is best illustrated in photo 2, page 176.)

The ⅛-in. wide masking tape is excellent for masking separation stripes where a base or intermediate color is covered by tape, then top coats applied. When the ⅛-in. tape is pulled off, a broad pinstripe remains. Because of its narrow width, it is also useful for masking off tight radius curves. Where ends meet, burnish edge to prevent paint from creeping underneath and cut off excess on outer edge with X-acto knife or single-edge razor blade with light touch to avoid cutting into paint.

THE PROPER COVERUP

After a design is established with narrow tape, adjacent areas must be protected from overspray. The best technique is to paper tape (with the masking paper tape dispenser) as close to the pattern as possible by applying the ¾-in., tape on top of the outline tape and filling in smaller areas with wider strips overlaping each other and the ends too (trim them with knife or razor blade). If done correctly, pulling the masking paper will also strip off the outline tape and filler pieces as well. Remember to put a loop of tape well into cracks (hood and trunk seams as well as door jambs) to minimize overspray. Pull all masking tape off immediately after top coat is shot. This lessens the chance of pulling up the newly sprayed coat with the tape. There will be less likelihood of harming the paint if the tape is pulled back over itself and slightly away from the painted area during removal. Should some of the tape's adhesive be left on the surface, it can be wiped off with PrepSol or enamel reducer. This shouldn't happen if the vehicle was prepped and painted right after being masked and said work wasn't done outdoors in direct sunlight.

NORIDGE TAPE

Reverse taping is a valuable technique for preventing solvent bleed-through onto areas covered

1. A painter refinishing a panel? Yes, but where would that painter be without tape? The masking tape that covers protruding brightwork, hides windows and wheels, and defines the area to be painted is one of the most useful products available to the painter.

2. While ¾-in. wide masking tape will handle most commercial auto painting chores, ½-in. and ¼-in. are most valuable too. The latter is used here to mask delicate script, though such ornamentation should be removed if at all possible.

3. The ¼-in. width is also a custom painter's best friend. It's most commonly used to outline designs because it can be formed around a very tight radius.

4. Once the ¼-in. defines the outline, it's back to the wider stuff to fill in small areas. Use ¼-in. tape as backstop.

5. Masking paper is dispensed with masking tape already affixed to one side, but it must be taped all around. Otherwise, air pressure from spray gun could blow free end over into wet paint.

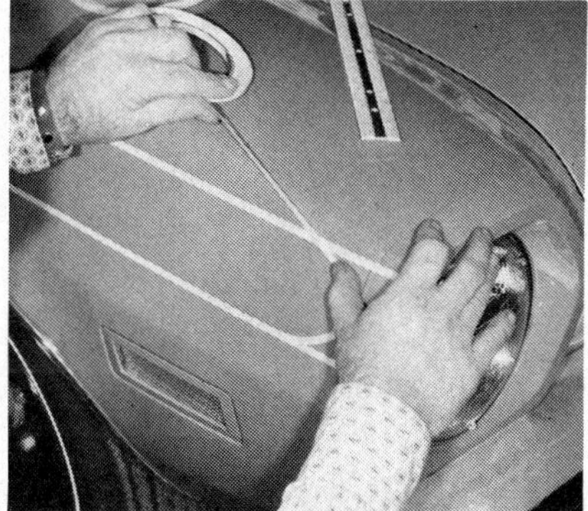

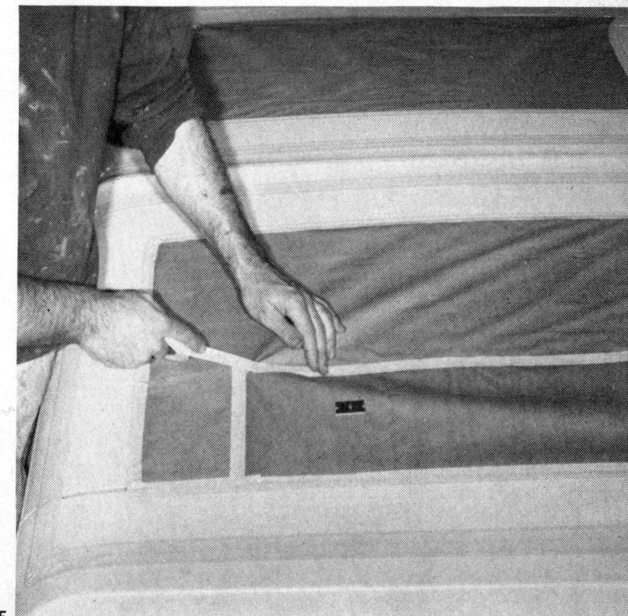

by masking paper. It is merely the technique of applying the taped masking paper backwards and folding it over so the tape on the paint and on the paper is on the underside (reverse). This makes the paper bow slightly and keeps it away from the body surface. This technique is handy, too, when covering areas where fresh paint is only surface dry. Don't confuse reverse taping with the technique known as back taping. Back taping is used to soften the edge when repainting a collision repaired panel. Back taping requires lifting ¼-in. of the edge of the masking tape at a slight (30°) angle on the side of the panel to be painted and directing the spray under it. The result of back taping is a smooth blending of the paint at the edge of the panel instead of the ever-present paint build-up along the edge of the tape that leaves a ridge when the tape is removed.

The problem in back taping is in lifting that sticky edge of masking tape away and keeping it away when subjected to the blast of the spray gun. The adhesive must be made a little less tacky and a nice smooth lift made from end to end. Very difficult task on long straight-line pulls until G.S. Staunton & Company, Inc., developed No-Ridge Tape. This is a specially developed ¾-in. wide masking tape that's zone coated with ½-in. of adhesive, leaving ¼-in. of non-adhesive area. When masking for panel repairs, two-tone finishes (including paneling, fogging, etc.) the non-adhesive edge is applied toward the area to be painted and turned up at a 30°angle to form a pocket. Paint sprayed into the pocket bounces back onto the panel and eliminates a build-up of paint along the edge. Hence, there's no rub-out or pinstriping needed because there's no unsightly ridge. This NoRidge Tape is a key element in Chrysler's patented "Sectioning Method" of paint repair which was developed by Charles Stephens of Chrysler Corp. This Sectioning Method is used exclusively in Chrysler, Dodge and Plymouth dealer's body shops and has resulted in better blending of repaint jobs and savings to the customers.

Before leaving the subject of masking tape, there are a couple of tricks worth passing on. When removing tape from an intricate job, use the point of a scriber to lift an edge next to the fresh paint. Fingernails are handy, but invariably leave their mark in the new finish. The second trick is an emergency measure should the tape begin to lift the film of fresh paint with it during removal. This sometimes happens when shooting complex custom paint jobs due to the length of time the masking tape is left on. Just dilute clear with thinner and spray it along the tape. The thinner will soften the paint and allow the tape to be removed without lifting paint too.

DECORATIVE TAPE

Perhaps the most obvious use for tape is in decorating a vehicle. The

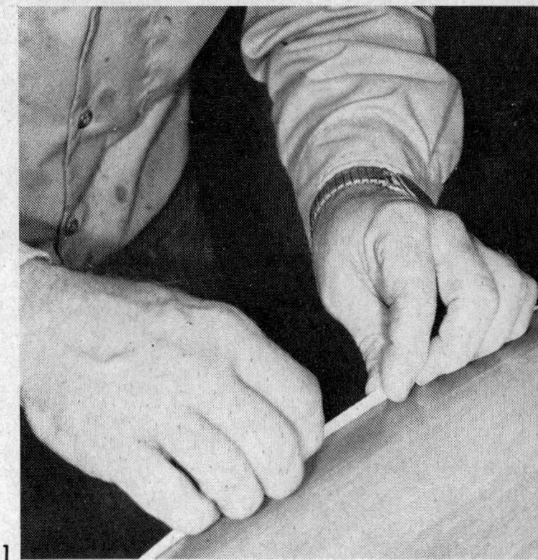

1

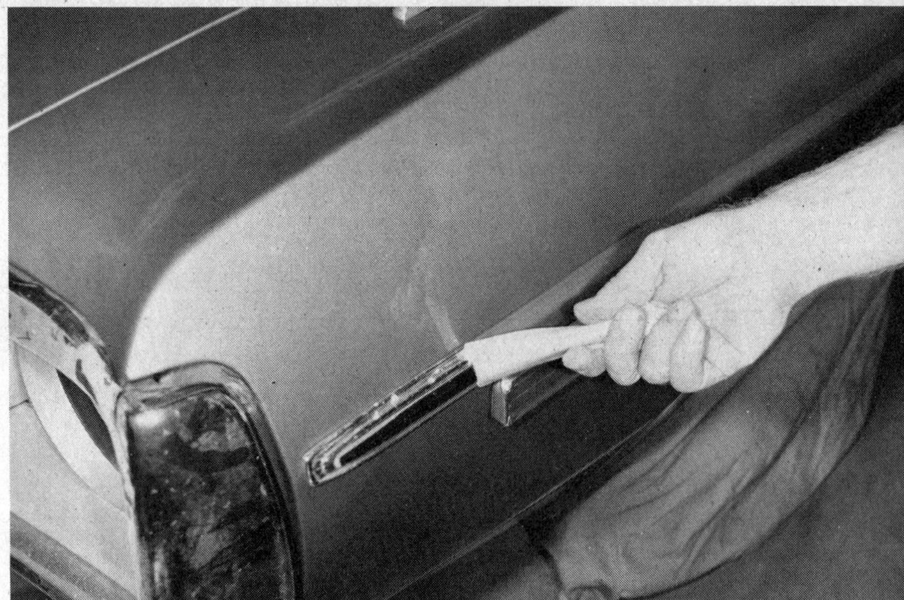

2

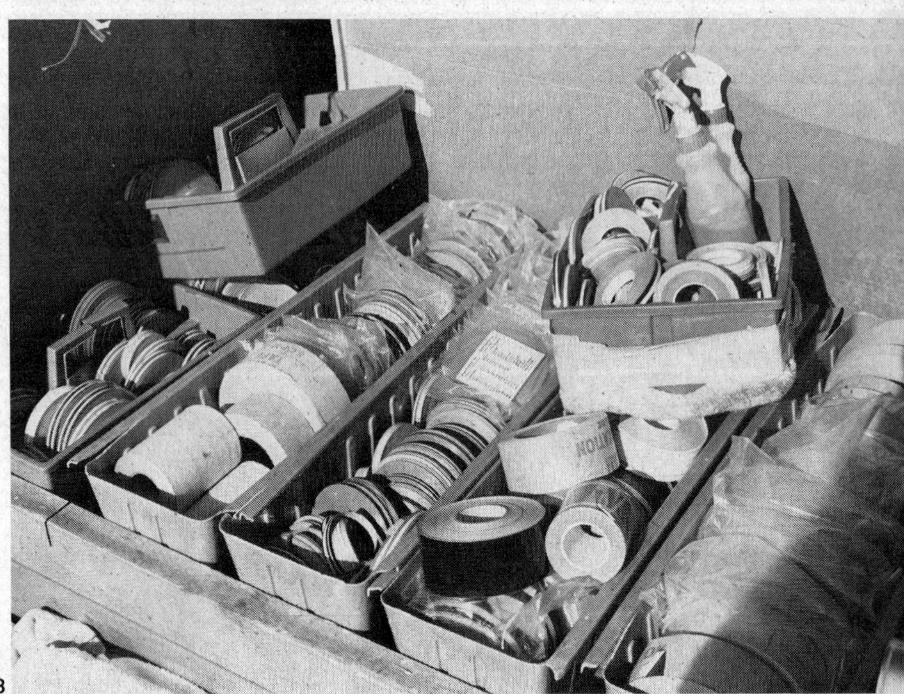

3

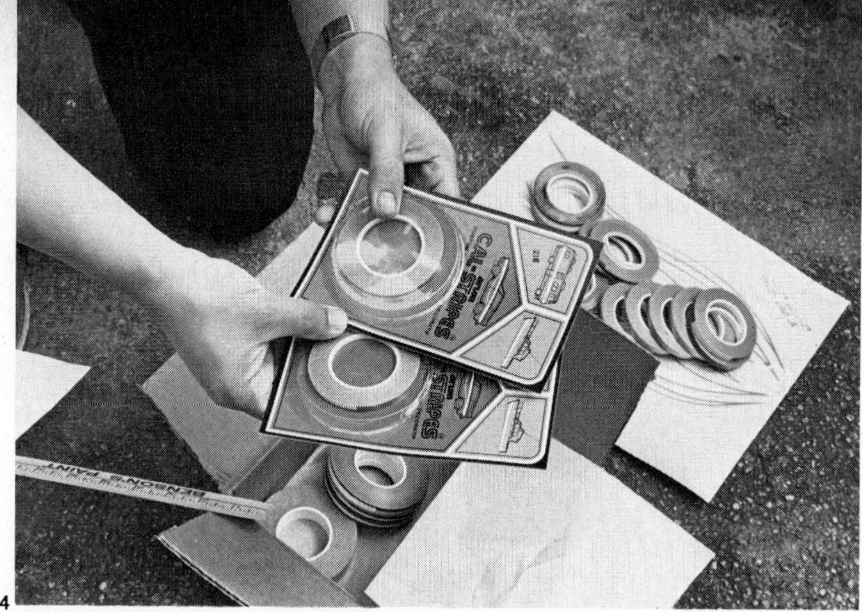

4

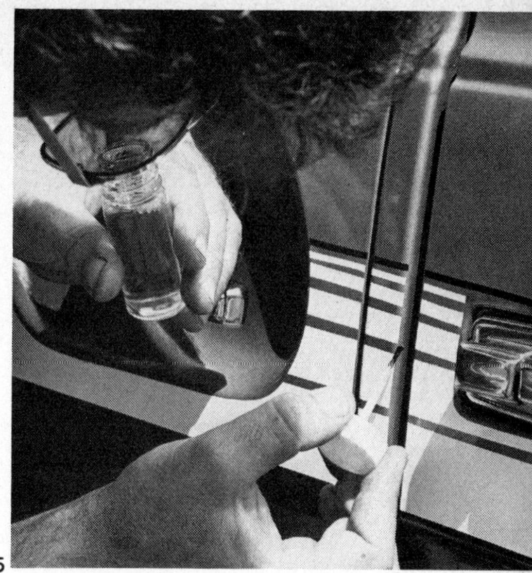

5

6

1. Reverse taping, where taped edge of masking paper is applied facing the wrong way and paper folded over tape seam, and back taping (shown), where edge of tape is lifted to form pocket and soften paint edge, are very useful techniques for custom painting. NoRidge Tape from G.S. Staunton & Co., Inc., makes back taping quick and easy.

2. When removing masking tape, it should be pulled away from the painted area and back over itself.

3. A peek into a professional tape striper's equipment locker should convince unbelievers that any kind of line can be laid. These rolls of 3M tape are for use by the pros.

4. Arlon Cal-Stripes are available in bulk rolls (for the pros) and blister packed in smaller quantities (for those one-shot jobs). Beware of low-priced variety store imitations. They may fall off or bleach out with passing of time.

5. Seal edges of tape stripes with Duralac (properly thinned to brushing consistency), not fingernail polish which will soon yellow. This graduated size, five stripe pattern comes from 3M. Black stripe above and below are separate.

6. Never run stripes behind door handles where fingernails might scratch and snag them. Note better appearance obtained by not wrapping tape around edges of door panels.

rapid development of super-thin, weather resistant plastic films and a corresponding advance in adhesives have made possible the staggering array of decorative tape products available today. But the real key to this cornucopia of vinyl, polyester, mylar tape was not in combining the plastic films and space-age adhesives, but in putting them in a form that made application quick and easy. From embellishments along rocker panels in the form of marque logos and strobe stripes to utilization virtually everywhere on a vehicle was a hard fought battle with design variety, quality and ease of application contributing to the victory. Spartan Plastics' rocker panel put-ons have begat a wild generation of Trimbrite goodies, and SED Products put diffraction tape in the spotlight with their quality polyester Space Tape and Streaker Stripes. Many companies fulfilled the deman for top line vinyl pinstripes (Arlon Cal-Stripes and the aforementioned companies among them) and this application has really caught hold. Then, too, bold graphics (four different colored stripes on the same sheet) as made by 3M Company have advanced the cause of vinyl on metal.

All this has given rise to a new profession. Joining the pinstripers and painters are the "custom tapers." To join this growing legion requires, in addition to artistic flair, a knowledge of these unique materials and what can and can't be done with them. How curves can be forced with a gentle application of heat; how to handle long pulls and large sheets; how application of heat or solvent or both can be used to remove stripes; and how to create astounding visual effects by taking advantage of the basic properties and incredible varieties of these materials must be learned before tackling that first commercial job on a stranger's vehicle.

BASIC TAPE DESIGNING

It is said that artistic flair is the ability to look at a plain object and visualize every step necessary to turn that object into a thing of beauty. In the case of a car, a truck or van, this vision might be a few simple stripes, a harmonious color treatment, or whatever. There are, in fact pinstripers, painters and, yes, tapers whose heads swim with wildly original ideas the moment they see a vehicle drive up and park. Their biggest problem is in selecting just one of those many ideas. Those not gifted with this artistic flair should probably content themselves with something else.

The desire to create is strong, however (even within those who have no gift for it), so the easy way out for the do-it-yourselfers is to practice designing on a picture. Take photographs of the vehicle, front, rear, top and sides, then have 8X10-in. enlargements (color or black & white) made. Tape these to a drawing board, kitchen table, etc., and tape a sheet of clear acetate over them. Draw in the customizing and pinstriping ideas with grease pencils or sharp crayons (it can be washed off with rubber cement thinner). This is a lot less expensive than wasting $20 to $50 worth of tape and finding out the design isn't pleasing.

The simpler the design, the easier it is going to be to apply. Curliques and baroque scrolls may look sensational, but can tie fingers into knots during applications. Besides, such things may put a lot of emphasis into design details that the factory stylists wanted to conceal. Items like door handles, hinges, gas filler doors and the like are there because they have to be, not because the stylists thought they would enhance the lines of the vehicle. Emphasizing them with striping may make these styling stumbling blocks more conspicuous than they should be. Think simple, think clean and it's surprising how much a good striping job can add.

Don't feel that the design on paper must be adhered to rigidly. There is a certain amount lost in the translation from two dimensions to three dimensions. When viewed from angles other than shown and used, the design thought ideal on the acetate may require modification or even rework. Don't despair, use that artistic license and alter the plan. Don't be stumped by the discovery that vehicles are not symmetrical. If one molding seems a bit higher, compensate. The saving feature here is the fact that the two sides can seldom be viewed at the same time (unless parking in front of highly reflective store windows).

Once the design is selected, it's time to get out the tape and go to it. There are a great many brand names on the market, but experience will dictate that price is in direct proportion to quality 99 out of 100 times, and the place to get quality tape is from suppliers to the automotive trade. Many of those offered for sale in drug stores and supermarkets will come off after exposure to extreme weather, or lose their lustre and turn dull, or bleach out after day-in, day-out exposure to the sun. Shortcut these problems by picking your brand from those stocked by automotive paint and refinishing supply jobbers and dealers.

As with custom painting, a good custom taping job can only be done on a surface free of dirt, silicone, wax and polish. The stripes are meant to be applied to paint and become almost a part of it. So after a thorough washing with mild detergent, use a stripper or degreaser (PrepSol). The surface, in addition to being squeeky clean, must be out of the sun and at proper temperature since the adhesive is thermosetting and time must be allowed to work with the tape before it begins setting. A good compromise is to try and maintain a surface temperature of 65°F., this will give about 12 hours in which to correct any mistakes or make changes in the design. Use of a sponge and a bucket of cold water in conjunction with a commercial heat gun (not a hair dryer) will make maintaining a good working temperature somewhat easier.

Before starting to stripe with tape, remove mirrors and any other chrome adornments that may be in the way. If any of these are to be embraced by the design, fine, leave them on—otherwise remove them to permit freedom of move-

1

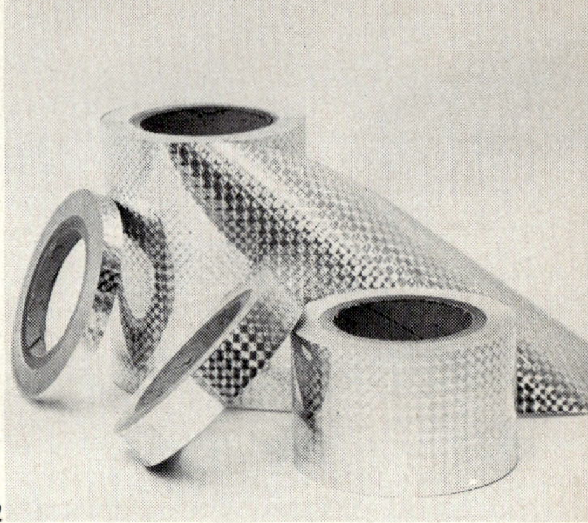
2

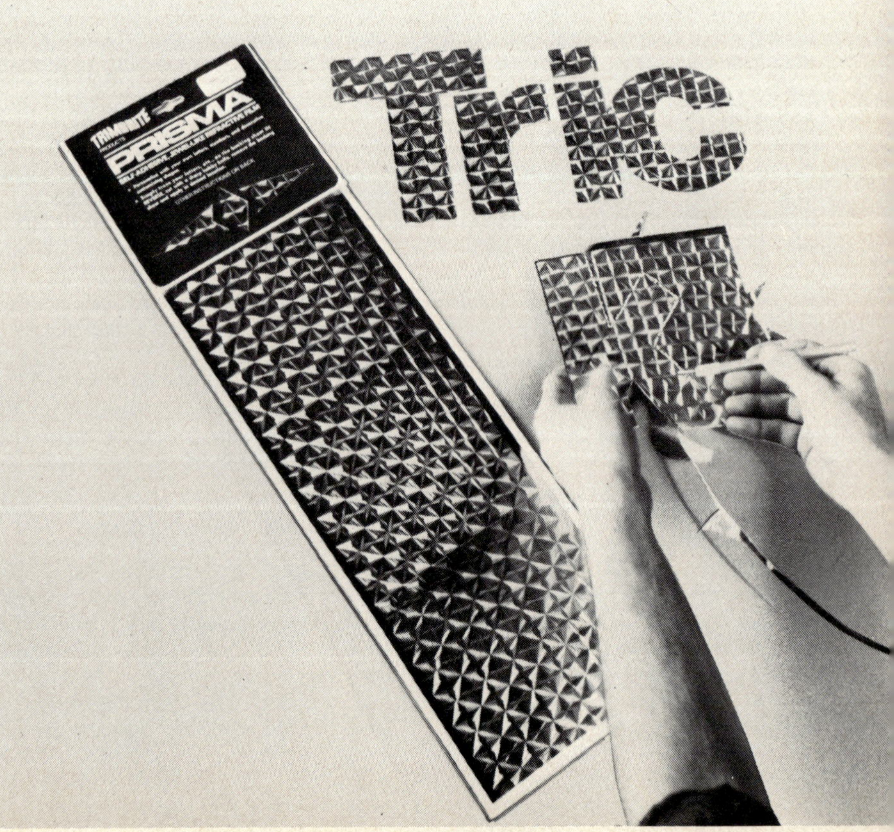

3

ment around and along the vehicle.

Now to discuss some helpful little tips and reveal some professional tricks. When pulling a straight line down the side of a vehicle, attach one end and strip off enough tape to run the whole stripe, but keep it slightly away from the panel. Squat down and sight along the panel. Put tension on the free end of the tape and bring it straight into the panel. Check straightness by eyeballing down the panel. If not satisfied, lift the end straight away from the panel and try again. When pleased with the alignment, walk back along the panel and pat the tape gently into contact with the panel. Sight again, and if satisfied that misalignment hasn't occured and the tape is touching along its entire length, gently rub it down with a soft cloth. Now about door seams and other panel breaks: don't cut the tape in the middle of the seam and wrap it around! Cut it right at the break in the panel (taking a section out of the tape equal to the width of the seam), burnish the ends down and coat with thinned Duralac (fingernail polish tends to turn yellow with age).

A single-edge razor blade is useful for cutting the tape and trimming ends, etc., but an X-Acto handle and a pack of No. 11 or 16 blades makes more sense when it comes to tape striping. Whichever instrument is selected, remember to exercise extreme care when cutting. Don't take the knife or razor and simply cut the tape across its width. This can also make a neat cut into the paint job, which can shortly turn into a not-so-neat rust spot. Whenever possible, use the blade as an edge to trim the tape against. Sever the free end with a quick snap. Tapering the ends of the tape makes for a far more professional appearance than merely whacking it off at right angles. Fortunately, the quality-made tape is thin enough that lines can be continued by overlaping the old end with the fresh end. In butting ends together though, use the tape beneath as a backstop for the blade. Don't forget to peel off the clear overlay before butting or overlapping ends. Needless to say, the piece facing forward should be overlapped, otherwise wind, weather, and who-knows-what can conspire to lift it.

Curving the tape, wide or narrow, is a technique that becomes second nature with practice. Experience will show, for example, how tight a radius can be formed using a particular brand and width of tape. For beginners, a couple of hints are in order. First of all, remember that the wider the tape, the broader radius a curve must be laid to prevent wrinkles. Second, turning a curve without putting wrinkles into the tape is a two-handed technique. One hand should be holding the roll of tape and putting tension on the stripe being laid onto the surface, while the other is actually pressing and smoothing and guiding the tape into the desired radius. On multi-stripe rolls, the tape is sandwiched between two peel-away protective sheets. Removing the bottom sheet

---

1. Treat tape ends with imagination instead of chopping them bluntly off. These ends have been given a fine taper with a knife blade.

2. Metalflake Incorporated's Beta tape comes in sheets or several roll sizes, and in three styles. This is Spectrim tape, more commonly known as diffraction tape.

3. Trimbrite's Prisma tape and SED Products' Space Tape are diffraction tapes. Space tape is offered in silver, gold, seven brilliant colors, even in pre-cut letters, numbers and popular logos.

4. Joe Simiyoshi created this superb portrait in Space Tape, by taking advantage of the variety of colors in the SED Products line. How about using it for van murals?

5. SED Products' Streaker Stripes are pre-cut pattern stripes of Space Tape. Some of the design possibilities are shown in center. These Streaker Stripes have clear overlay that is peeled off after Space Tape pieces are positioned.

4

5

in the sandwich exposes the adhesive (either acrylic or rubber-based —with the former being preferable, of course). The clear top sheet of the sandwich holds the two or three (or more) stripes in perfect alignment and should only be removed by peeling it straight back over itself. Before removing this clear top sheet, check to make sure the stripes are as desired because when the top sheet is removed it is extremely difficult to lift the stripes and reposition them properly. This clear film can be cut and peeled away to allow realigning individual stripes in curves (very delicate business) or when tapering ends, of course. Never go directly from a straight run into a curve. Take a truck driver's swing. That is, curve the tape ever so slightly in the direction opposite the curve just before entering it. Likewise, if the radius is being pulled over a compound curve, it will be necessary to slightly shorten the radius of the inside stripe(s) or the illusion will be that the parallel stripes are closer at the apex. When going from the back of a vehicle around to the sides with stripes, be sure to raise them just a bit or the illusion will make them appear to drop. Vehicles with curved flanks are also guaranteed to drive the novice tape striper up a wall. Best solution is to measure and mark before running a long stripe. As mentioned before, practice and experience will make dealing with problems like these second nature.

Sooner or later (more likely sooner) the custom taper is going to be forced to handle those wide, unwieldy graphic panels of multi-color stripes, textures and/or patterns. Skill at handling these marks the real craftsman. Preparation is the same (the need for a truly clean surface is universal), but the application is totally different. The location of the vinyl or polyester panel should be marked with a Stabilo pencil. An alternate is to affix strips of masking tape that hang into the area where the panel is to be placed and make the guide marks on the masking tape. These pieces of tape can then be used to hold the panel in position while the surface is washed down with wetting solution and the backing peeled away from the vinyl sheet or panel.

The washdown with wetting solution is the secret to a good smooth application. If the surface were dry and the backing removed, the adhesive would bond the vinyl on contact wherever it happened to touch. By using the wetting solution, the vinyl with exposed adhesive rides on the wet film and can be adjusted, repositioned and generally played with until perfect alignment is obtained. Pros, of course, have their own secret solutions and mixtures, but among the best is a few drops of dishwashing detergent in a bucket of plain water. This is applied with a sponge until the entire area is covered with a film of soapy water. Then the backing is removed from the

1

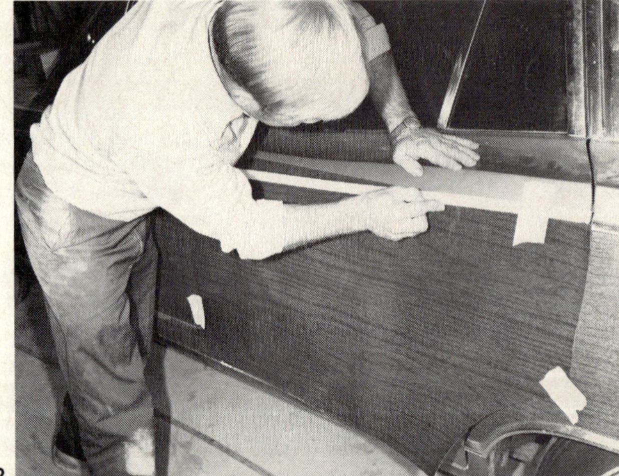
2

3

4

1. Working with broad vinyl sheets requires special technique. This series shows wood-grain trim replacement on station wagon, but application is the same for any type vinyl tape or sheet. Use tape to temporarily secure vinyl before trimming a piece to rough size.

2. If pattern is involved, be sure of alignment. Upper trim line is marked out in masking tape by feeling for clip holes that secure trim piece. Overlap must be allowed for in this application.

3. Cleaned surface is given a liberal coating of soapy water with lint-free rag or sponge. Some pros prefer to use their own wetting solution mixture, but mild detergent in water is a satisfactory answer.

4. Carefully peel off the backing that protects adhesive. Avoid stretching film or kinking vinyl and remember, it's really sticky.

5. Line up previously marked edge and gently place vinyl in contact with surface. If solution has been applied correctly, the vinyl sheet will be able to be slid around on surface until perfect alignment is obtained. Guide marks are helpful.

6. Taking great care not to move the newly applied vinyl overlay, squeegee the water out from under the sheet with rubber Bondo pad. Work from center outward pushing air pockets and bubbles toward edge.

7. Heat gun may be required to get vinyl to conform to compound curves or to stretch around panel edges. Heat also triggers adhesive and speeds setup time.

8. Metalflake's new Spray Mask may signal the beginning of new era in elaborate custom paint jobs.

graphic panel and it is laid gently on top of the water film. When adjusted so it's placed right, the water is squeegeed out, beginning in the center and working toward the edges. Most air bubble can be worked out to the edge with a little care and great patience, but stubborn ones can be punctured and the air squeezed out through the hole. The hole is invisible in the finished job.

Since panels to which the vinyl sheets are applied are seldom flat, a certain amount of stretching is required for a perfect job. The use of an industrial-quality heat gun (similar to a hair drier, but more robust) will prove to be a great help—it will also assist in speeding up evaporation of the remaining wetting solution.

Diffraction tape is one of the latest additons to the ever growing variety of decorative tapes—although it is also sold in sheets and rolls. The best choice here is polyester material and tops among the manufacturers of diffraction tape using this material are SED Products and Metalflake Inc. The newest of Metalflake's Beta tapes is 3-D. This is not a diffraction tape that breaks light up into a rainbow. Rather, 3-D conveys the illusion of depth. When a stripe of 3-D is applied to the hood of a car, it looks to the viewer as though it is ¼-in. beneath the surface on which it's applied. Metalflake's Spectrim is the light diffracting tape that changes colors as the angle of view changes. SED Products offers their highly popular Space Tape (another true diffraction tape) in the standard silver as well as gold, royal blue, orange, green, bronze, light blue, red and black walnut; thus opening up unlimited possibilities for a creative taper. SED Products' newest item combines the visually arresting pattern and attractive colors with a pre-cut design in roll form.

Application of 3-D, Spectrim, Galaxi, and Space Tape is similar to vinyl striping tape, but curves and contours must be less radical. There is a protective backing that must be removed to expose the adhesive, but there is no peel-off top coat.

Absolutely astounding effects are possible with tape, whether it's mind-boggling endless line patterns utilizing masking tape or the ever-changing rainbow of diffraction tape. However, the real trick is to master the medium and go on from there to explore new patterns or seek new potential.

Metalflake Incorporated's new water-based sprayable coating for custom painting deserves mention, too. Sold under the name Metalflake Spray Mask, this product may well revolutionize the art of custom painting. The milky liquid sprays or brushes on then dries transparent. It can be easily cut with an X-Acto knife or razor blade and peeled off the area to be painted. Since it does not interact with automotive paints, it's superb for intricate designs and overlays that would be exceptionally difficult to duplicate with masking tape. But don't give up on tape—it's still a mighty handy (and decorative) medium.

8

5

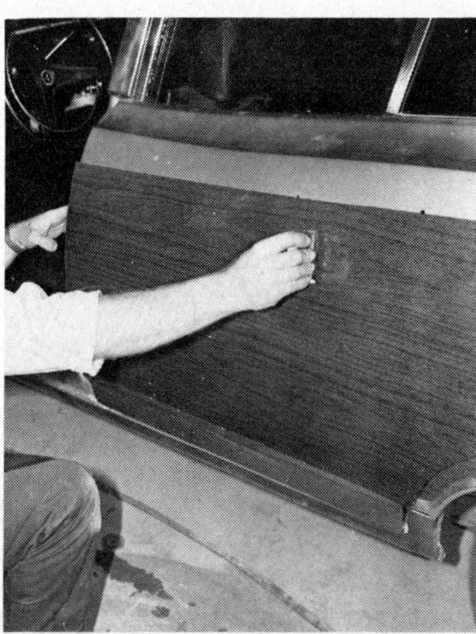
6

7

How To Apply Candies, Pearls and Flakes

Each type and brand of custom paint requires a technique of its own.

by Jay Storer

There are some painters who admire custom paint jobs but are afraid to tackle them. They assume that candies, pearls and flakes are just too difficult for anyone but a specialist to take on, but it's our contention that anyone with a full mastery of standard automotive painting can become a custom painter with a little steering and a few practice sessions.

We're going to assume that readers of this book already have a firm foundation in the basics of spray equipment selection, the differences in application and mixing for the three basic paints (lacquer, acrylic, and enamel), and have a good spray pattern mastered. All of the same techniques and tips that cover standard paint jobs will apply to custom work though they sometimes differ in degree.

Custom painting differs mainly in that this type of work is much less forgiving of any glitch or miscalculation than when using standard opaque colors. For example, your preparation has to be good for a standard paint job, for a custom job it has to be flawless. When you're dealing with custom paints, many of which are translucent, you can't sand out runs, bugs or other defects without ruining the effects. Taping and masking have to be perfect for a custom job. For an

1

2

1. Variety and creativity in custom painting has to start with materials, and SEM Products, makers of the line of Aero-Lac paints, has 'em. A full line of flakes, pearls, and candies are available in bulk or in spray cans.

2. Many custom painters use the term "Murano" as a generic term for pearl pigment, but this is only one line of pearl products made by the Mearl Corporation. At left is one of their powdered pearls (Flamenco) which can be mixed with any paint. Center here are two Murano jars of colored pearls, and at right are two jars of straight or "mother of pearl white," which comes in several stages of brilliance.

3. This chart from Aero-Lac shows the various "cast" results from using the five basic colors of pearl pigments. The white base is by far the most popular for custom pearl painting.

4. One of the originators of custom paints, Metalflake, Inc. has a full line of respected products for trick painting. Some of their products are available from no other source.

5. Metalflake also markets their own airbrushes and sprayguns, including this special self-agitating gun for use with flakes and pearls.

6. Stewart-Warner's Alemite Division also makes a self-agitating spraygun for use with metallic and other paints with heavy pigments. A slight amount of air pressure is bled off to drive a fan that makes the plastic-covered magnet (arrow) spin in the cup.

ordinary paint job you just have to keep the paint off of the chrome and glass and a little overspray can be cleaned off later; not so with a custom job where intricate masking between design panels leaves no room for error, since you are masking one paint from another and you can't use thinner to erase any overspray. Above all, when using custom paints, your spray pattern must be exacting to deposit even layers of paint on all surfaces or the trick effect of whatever custom paint you're using will be mottled.

There are three basic types of custom finishes, candies, pearls and flakes, although there are diffences between manufacturers of these types, and custom variations of the basic three. None of them are one-step or one-coat processes, and you can expect to have the vehicle in your shop or paint booth for a lot longer than a comparable "straight" paint job. You'll be painting as many as three different types of paint on top of each other, and several coats of each one. With proper drying times between all of these coats, plus remasking and respraying when doing a paneling job, mural or other multicolor effect, the subject of some custom paint jobs may be in the shop for as long as a week or more.

Because of the ease of repair, quick drying time, availability and durability (because of the thinness of the coats) lacquer is the most often used medium for custom painting. When we say lacquer, we mean acrylic lacquer, since most paint manufacturers are phasing out their straight lacquer lines in favor of the acrylics whose liquid plastic base makes for greater durability. The exception to the rule-like use of acrylic lacquers for custom painting is in clears. Lacquers don't build up a surface as quickly as enamels, which is why they are used in multi-layer custom jobs to keep the overall job from getting too thick (which leads to cracking), but this can be a drawback when your trick stuff is done and you're trying to cover it completely with clear for depth and protection. That's a point at which a thicker paint can be helpful and time-saving (since less coats are required), and often the job is covered with epoxy or polyurethane clear, especially on race cars where the fuel-proof protection of the two-part plastic paints is an added bonus.

FLAKE PAINTING

Like Kleenex, Metalflake has become a generic term for flake paints, although it is a trademarked name of a particular company. The Metalflake Company originated the idea of using large metallic flakes (larger than the particles in a standard metallic paint) suspended in clear back in 1961 and they are the largest producer of such paints. However, a good idea is hard to keep to yourself, and now there are other paint companies with similar products. In this chapter, when we mention "flake" painting, we are referring to any of the flake paints, and when we mention Metalflake with a capital M, we mean a product of Metalflake Incorporated This may be confusing to some because Metalflake Inc. also makes their own candies and pearls. The other major custom paint manufacturer, SEM Products, has their own brand of flake paint called

PEARL COLOR	WHITE BASE	BLACK BASE
PLATINUM	Brilliant silvery white color. Used for true mother of pearl effect.	Lustrous silver color used in black pearl finishes.
GOLD	Golden white color that reflects a bluish cast in certain lights.	Brilliant gold color with extremely lustrous sparkle, used as a base for brilliant candy colors.
BLUE	Bluish white color with a golden cast.	Vibrant blue color with high luster.
RED	Reddish white effect with a green cast in reflected light.	Bright red color with brilliant luster, used for accent effects.
GREEN	Greenish white with a red reflection color.	Brilliant green color with high luster. Makes a beautiful deep green.

3

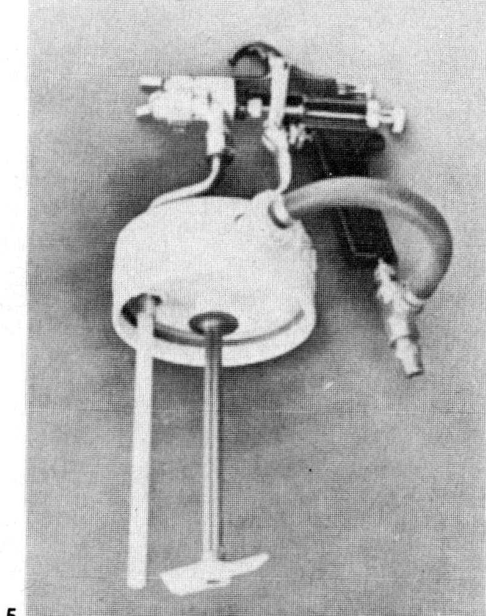

5

4

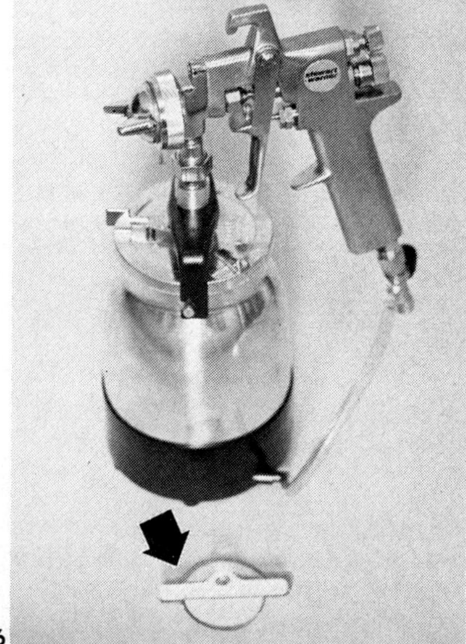

6

Aero-Flake, so when you see it referred to as such, we mean a product of SEM. Now that you're sufficiently confused, we can continue.

Although the original Metalflake was made up from actual tiny squares of metal foil, today all flake paints are made from pieces of Mylar which is just as brilliant but doesn't tarnish. The basic idea is that these flakes are sprayed onto a car while suspended in clear paint, and when the desired effect has been achieved and the clear is dry, more coats of clear are applied to even out the surface. The Mylar particles reflect light back through the topcoats, resulting in a shimmering surface in which there is no ordinary pigment, the flakes are the pigment.

Since it is almost impossible to cover a vehicle's skin completely with flakes and there will always be some minute spaces between the flakes, a ground or base coat is needed. The base coat is generally a complimentary color to the desired flake paint job. In other words, if you're going to apply a blue flake job on your vehicle, you need to lay down a blue base coat so the spaces between the flakes won't show primer. You can use any standard lacquer solid color for a base, or even a metallic color base, but you must apply a nice, even base coat with no streaks or light and dark spots. Metalflake Inc. has special colored primer-surfacers for use with their products which save extra steps in painting because once you have a good solid primer job on the car, the "background color" is already there and you don't have to apply a separate ground coat.

After the ground coat you're ready for the flakes. These come either as dry powders you mix with paint yourself, or they can be purchased pre-mixed. The advantage of not using the pre-mixed flake is that you can mix it with any color that you want. Metalflake Inc. Company and SEM Products have taken two different approaches to the flake business. Metalflake still makes their Mylar flakes in a variety of colors, that is the flakes themselves are colored. Thus if you want a "Shimmering Lake Blue" Metalflake paint job, you use that color flake from Metalflake's selection of 36 flake colors. The Metalflake flakes are mixed with clear acrylic lacquer for spraying, since no additional coloring is needed beyond the flakes and the base coat. SEM's approach is to offer a plain silver flake, which is used with various candy colors instead of clear. The resulting paint job color is achieved through the candy vehicle for the flakes, since the light reflected from the silver flakes comes through the translucent candy. Both arrive at similar results, the main difference from the custom painter's standpoint is a stocking problem. With one manufacturer's line, he stocks many colors of flakes and only one vehicle color, clear. With the other line, he stocks

1

2

3

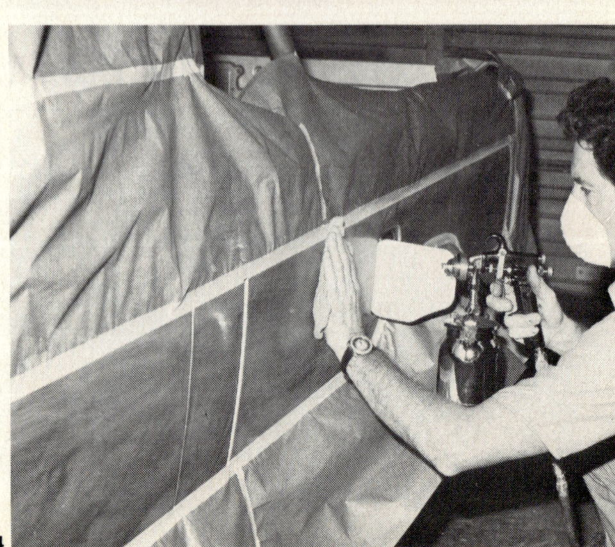
4

only one color of flake, silver, but stocks a variety of candy colors as vehicles for the flake.

The biggest challenge of doing a flake paint job is getting the tiny pieces of Mylar or polyester to lay down. Static electricity can be a real problem with flake, especially on a fiberglass car like a Corvette. Whether metal or fiberglass, it's advisable to attach a steel chain to some part of the car's frame or suspension and let the other end touch the ground. Otherwise, the static you build into a car when you constantly rub it back and forth with sandpaper or tack rags may cause the flakes to stand on end. This is one of the flake painter's mortal fears, because if some of the flakes stand on end, it will take a zillion coats of clear to make the surface smooth. One of the secrets of applying clear or candy with flakes suspended in it is to thin the medium (clear) down considerably and use a lower-than-normal gun pressure, like 25 lbs. Too much air pressure and the flakes will bounce off the car. For an even effect, it's important to crisscross your consecutive coats. Spray one coat in a vertical plane, the next horizontally, and don't start and stop your gun in the same places every time. You want the flakes evenly distributed over all of your base coat. Since the flakes are considerably larger than the particles in metallic paint, they have a tendency to fall out of suspension in the clear or candy, so it's important to keep the gun thoroughly agitated during the painting process. Give the cup a shake after every pass. In the past, painters have resorted to using all sorts of objects in their paint cups to aid agitation, ball bearings, marbles, even nuts and bolts. These help, but you still have to maintain frequent agitation. Something new that has come along is the self-agitating spraygun. This is a boon to sprayers of standard metallic paints as well as custom painters, affording more uniform results and there is no danger of spilling paint out of the bleed hole because of over-active hand agitation. Metalflake Inc., markets one of these guns, which they call the "Equalizer", sort of a Colt 45 for painters. Another self-agitating gun is made by the Alemite Division of Stewart-Warner which is slightly different in construction. Both guns bleed off a little of the air pressure just before it enters the gun handle. In the Metalflake design, this air drives a small fan connected to a shaft that sticks down into the paint cup and has a mini-propellor attached to agitate the paint. The Alemite gun channels the air to a housing under the paint cup instead of above it, where it drives a metal fan. Inside the cup is a plastic-covered (for ease of cleaning) magnetic spinner that follows the fan without a physical connection between them.

Despite all of your efforts to the contrary, some of the flakes will not lay down completely. It's part of the nature and appeal of a flake job that the flakes lie at all sorts of angles so that the paint job sparkles from many viewpoints. Thus, you have to spray something over the flake to cover them and provide a smooth, glossy surface with depth. Clear acrylic lacquer is generally used, although you can also use acrylic enamel clear, or a two-part epoxy or polyurethane clear. With the standard clear acrylic lacquer, you'll have to spray several coats, wet-sand with #400 paper, then spray some more clear. This is repeated until the surface is uniformly smooth. Using the thicker clears such as the epoxies and polyurethanes will take fewer coats, but you have to wait a lot longer between coats to sand the two-part paints. With standard acrylic lacquer clear, you need only wait about an hour (or whatever the manufacturer recommends) before wet-sanding, while the two-part paints need to cure at least for 24 hours before they can be sanded. Now are you beginning to see why a custom paint job takes so long? The amount of coats it takes to cover the flakes and provide a flat, shiny surface depends not only on the spray technique you used to

1. Our custom painting exercise with Gene "The Man With the Golden Gun" Winfield began with a thorough rub with wax-and-grease remover.

2. Spend a lot of time on the masking of your trick paint, and don't press down the tape for good until you're satisfied that every line is what you want. Winfield is using ¼-in. tape here. We were planning on a band of various paints surrounded by a wide border of light candy blue.

3. After taping off the center part of our Courier's sides, the stock paint was roughed up with a ScotchBrite pad. Go lightly with this or use #400 wet-and-dry sandpaper instead. Get all the gloss off of the stock paint.

4. Winfield uses his gold-plated Binks to blow dust out of crevices while he tacks off the sanded paint.

5. The gun that usually paints a Rolls or a Stutz here applies black Aero-Lac lacquer as a ground coat for the pearlescent Design Colors.

6. We used all five of the Design Color pearls from Aero-Lac (SEM). All of them look milk-white in the can, but lots of agitation is necessary to keep the almost-invisible pearl pigments in suspension in the gun.

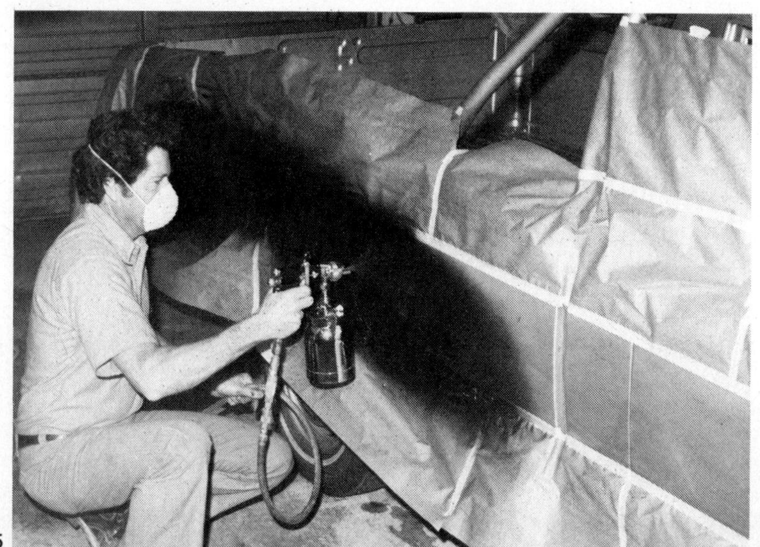

5

6

apply it, but also on the coarseness of the flakes. SEM's Aero-Flake comes in a fine or coarse grade, coarse having larger flakes, and the Metalflake line comes in four sizes. The more coarse the flake, the rougher the surface will be if some of the flakes stand up, and thus the more clear it will take to cover them.

CANDY PAINTING

Candy paint has been around since the early days of customizing, yet it remains one of the most attractive of custom finishes. Anyone who has even seen a good candy paint job can tell you how they thought they could walk right into the paint, and how even from 30 feet away the car seemed to actually glow. Candies are translucent colors, almost like straight clear with just a little pigmented toner added, and are applied over a reflective base. The number of coats applied determines the depth or density of the final color; it builds a little at a time.

No other custom finish can match the dazzling appeal of a good candy job, but part of its appeal is the fact that it is also the most difficult custom paint to apply properly. Candy is the one custom finish most admired by custom painters themselves, because when they see a good candy job they know how much time and talent went into it. Travel to a custom car show or gathering of street rods or street machines and the true mile-deep candy job will be conspicuous in its scarcity. By the same token, if you're a custom car owner, or a bodyshop operator looking to stimulate business, no other paint job will attract as much attention or showcase your talents as well as a candy job.

What gives a candy job its brilliance is its underbase. The candy itself really adds only a coloring, like spraying a thin-colored dye over a light bulb to change it from white light to red or blue light. In the case of a candy paint job, the reflective underbase plays the part of the light bulb, so you can see how important the base coat must be to the overall result. Traditionally, candy paints are applied over gold or silver base coats, but white can also be used. All three colors vary in their reflectance, but more importantly, in the "mood" or character they give to a candy job. The same candy color applied over different bases will appear slightly different in tone. The gold bases give the "warmest" tones, the silver is cooler, and the white base is cooler yet. You have to look at various paint chips or check the paint jobs at a large car show to really get an idea of the effects the base coats will have. When you see a candy job you really like, ask the owner or painter what he used as a base.

Whatever base coat you decide on, it must be applied evenly over the car. Flake paint may cover most of the blotches or errors in a base coat, but not candies. The metallic paints sold by the custom paint companies for use as candy base coats seem to have a lot more metallic particles in them than regular metallic lacquers, to give them more brilliance. Because of this, they need extra attention with regards to stirring or shaking before use. Even after thorough stirring there may be globs of metallic in the bottom of the paint can that you'll have to scoop up and disperse. Like flakes, this much metallic in a paint means it settles out quickly, and you have to shake the spraygun cup often to keep the metallic dispersed. If you don't, then parts of your underbase will be more reflective than others and the final candy job will look blotchy.

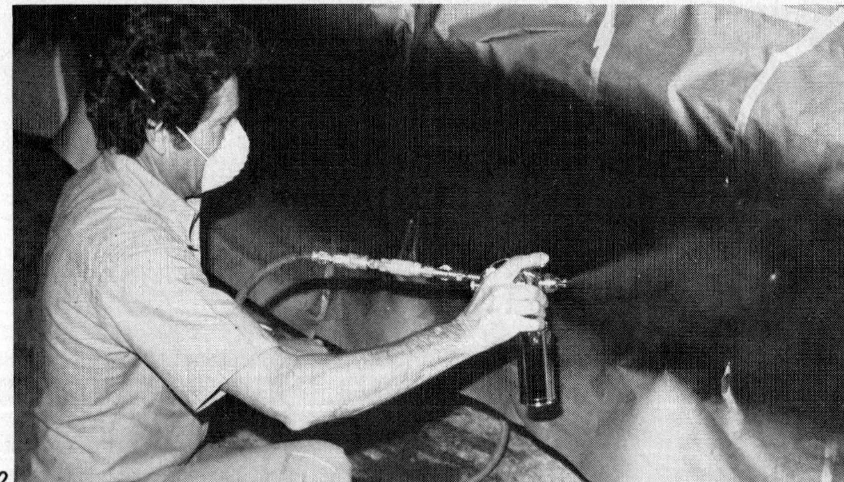

2

1

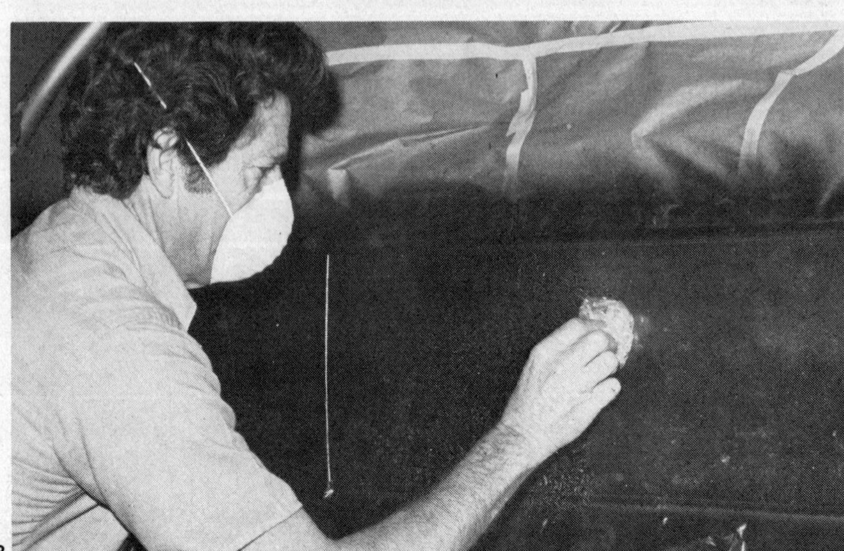

3

Generally, the metallic underbase paints don't have to be color-sanded before applying the candy. Your final clear coats will be color-sanded, so they will provide the flatness that makes a glossy job, and slight "orange peel" in the metallic base coat is desirable anyway because it means some of the metallic particles are standing at different angles and will make the final job more sparkly. You should lightly sand the base coat with #400 or #600 paper though, to get off any dust specs or other glitches in the base, because the candy coats will not cover any defects in the base. In fact, candy may call attention to any specks by gathering around a dust particle in the base and making a tiny dark spot in the finished job. Don't use any sandpaper coarser than #400, because the candy will also not cover any scratches you put into the base. A final caution on base coats is to avoid fingerprints if at all possible. The oils and acids on your fingers can cause invisible fingerprints that will tarnish the metallic base after the fingerprints dry. For this reason, some custom painters apply a light coat or two of clear over the base coat before sanding any glitches or proceeding with the candy.

With a perfect base coat applied, you're ready for the hard part, the candy coats. The acrylic lacquer candy colors are usually thinned with lacquer thinner (1 part candy to 1½-2 parts of thinner) according to the manufacturer's directions and local weather conditions, and sprayed over the reflective base. The hardest part of this phase is keeping a dead-even layer of candy over the entire surface. If you stop the spraygun twice in the same spot, that spot will be darker than the rest of the car, since the more candy you spray on the darker the job gets. Generally you have to keep spraying thin coats until you have the depth and color you want. Be advised though that if you spray too much candy on that the effect will be lost because you've cut down the reflectance of the underbase too much. The best advice we can give for getting the candy on evenly is to make each succeeding

4

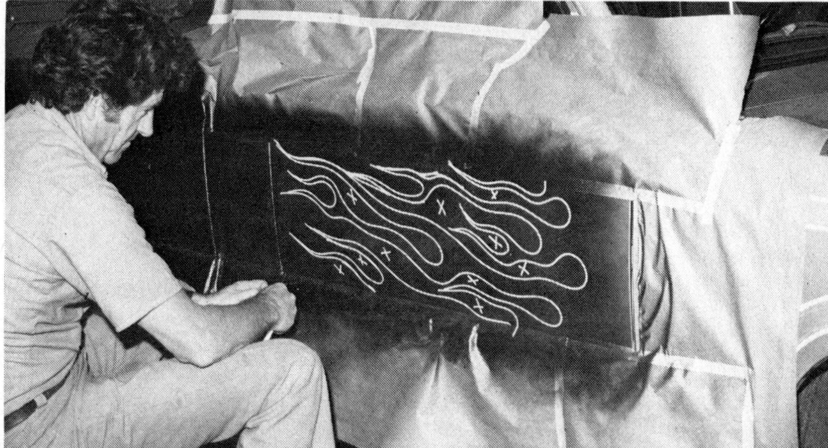

5

6

1. As with any heavy-pigment paint, it's possible for a tiny lump of the particles to clog your gun and spit onto the car, so even pearl paints should be strained with paint filter.

2. Using the smaller of his golden guns, Winfield sprays the first of the colors, silver. The "erasable" pearls look best over a dark base.

3. Design Colors can be brushed, patted with tinfoil, swirled with your fingers or patted, as here, with plastic wrap for various effects. Experience will tell you how much Design Color to spray on for each type of action. Mistakes are erasable.

4. Each of the colors was blended into the next Design Color, because they make different colors when put over one another. Ours went from silver to gold, red, blue and green.

5. Right on the doors, we decided to mask off some flames, using ⅛-in. masking tape. The X's here make it easier to see which parts are going to be masked off and which are flames.

6. The unwanted area is covered with various sizes of masking tape until only the flames are exposed. Color on the front fenders is stock silver.

coat in a different direction. Make one coat vertically, the next horizontally, then maybe a coat diagonally. Don't stop and start your spraying at the same line each time. Many painters are in the habit of stopping or starting their passes at the end of a fender, door or other panel, but in a candy job this will create dark lines or streaks, known as "railroad tracks" to custom painters. You must also remember that if you get any runs in your candy, then you'll have to start the whole paint job all over again! You cannot sand out the runs in a candy job because you can never respray it and get the same density. Since even a little run is a major disaster in a candy job, most custom painters spray their candy coats relatively thin and dry, almost "dusting" on the paint. You should wait at least 15 minutes between coats with candy, so that succeeding coats don't soften the previous coats enough to make a run. After the last candy coat is applied and you have the effect you want, you can relax for an hour or two and pat yourself on the back, the worst is over. After the last coats are thoroughly dry, you can apply normal clear coats to protect the candy. Two or three coats of acrylic lacquer clear or one coat of the two-part clears should do the job. As you know, the more paint you have on the surface, the greater the chance of the paint cracking later on due to the expansion and contraction of the metal during weather changes. The quandary of the candy painter is that the clear coats are what give the candy job it's depth and gloss. Only experience will tell you how many coats of clear can be safely applied without danger of cracking, because it also depends on how many coats of candy were applied and how thick the total paint job is. Your final clear coat can be color-sanded and buffed out after it has cured (see manufacturer's instructions for rub-out time of the paint your're using).

As you can imagine, a beautiful candy paint job is not arrived at overnight, and the total buildup of coats of primer, basecoat, candy and clear coats leads to a surface prone to cracking. Other drawbacks include easy fading in direct sunlight, and possible yellowing of the clear coats. The two-part epoxy and polyurethane clears are much more resistant to cracking and yellowing, but you still have a thick paint job and the best solution is to take care of the car. Some candy jobs that are neglected can be ruined in a year or two, and yet we've seen gorgeous candy jobs six or seven years old that have been kept clean, garaged and cared for.

A tip from customizer Carl Green addresses itself to two of the drawbacks of a candy job, the time involved and the number of coats of paint. Instead of shooting a gold or silver underbase, spray a base coat of a solid, opaque color that matches the final color you want for the candy. Color-sand this base, then shoot one or two coats of gold pearl. Over this, you need only apply a few coats of the desired candy color to achieve good results, so the total job takes less time and paint.

1

PEARL PAINTING

Our third favorite among custom painters are the pearls. There are a number of brand names and types of pearl pigments, some are claimed by their makers to be made from fish scales, some from abalone shells, and some from synthetic ingredients. All have slightly different effects, although they all share in common a hard-to-

2

3

describe pearlescent and irridescent quality. Pearls are the subtlest of custom paints, and their effect is only noticed when light strikes the car's surface a certain way, revealing a shimmering change of color on curves and highlights. For this reason, pearls are less suitable for boxy cars with straight, flat slab sides and flat hoods, and appear to their best advantage on compound curves. They are particular favorites as finishes for older cars with their more rounded fenders. Pearlescent paints have always been especially attractive as finishes for show cars, because the flourescent lights the shows are held under make the pearls even more brilliant.

Like the candies and flakes, pearls require a base coat. In this case, the base coat doesn't necessarily reflect light back through the pearl, but it does ensure that the voids between the pearl particles won't be noticeable. You can use just about any color for a base coat, even black which makes for some interesting and subtle effects. Commonly, the base coat is a straight-pigmented or non-metallic opaque color that is close to the color you want for the final job. This base is carefully color-sanded with #400 or #600 wet until it is flat and smooth with no orange peel. The pearlescent, which may be in a powdered or thick liquid concentrate form, is mixed with some clear acrylic lacquer and sprayed in light, dry passes over the base. The same cautions we mentioned about spraying candies apply to pearls. You cannot afford to have uneven light and dark spots and again, one run will ruin the entire job. When you look at pearls in the can, they look like thick milk, which is to say they look white. The more pearl you spray on, the greater the effect will be after it is cleared, but also too much pearl will leave the surface cloudy or milky and you won't get the proper effect at all. After successful application of the pearl, you let it dry to manufacturer's recommendations and then coat with just enough coats of clear to be able to color-sand and buff the clear without scratching through to the pearl, which would ruin your week completely.

As with candies and flakes, there are a variety of methods used in achieving pearl effects. In a standard pearl paint job, the pearl is mixed with clear for application, but custom painters often apply it suspended in candy instead of straight clear, which heightens the color effect, and it is also possible to mix pearl concentrates in standard metallic or non-metallic lacquers. This doesn't have the same impact as pearl by itself, but is more durable due to the lesser

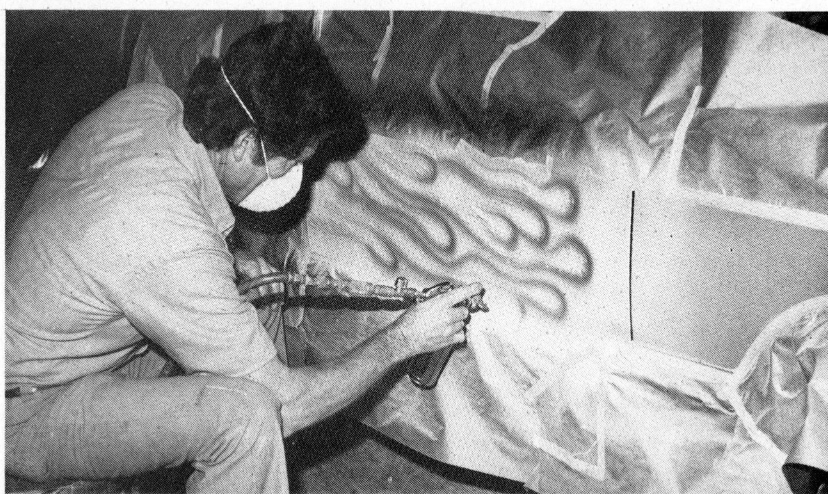
4

5

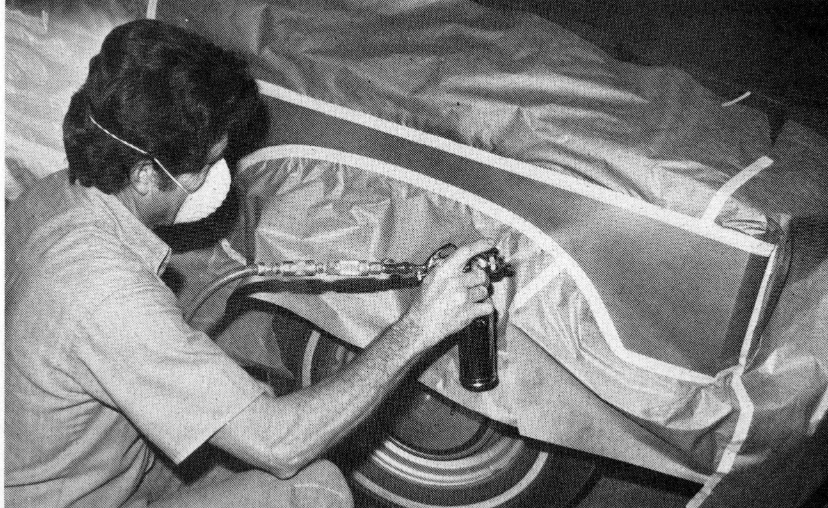
6

1. Since it would take too many coats of candy to cover our black base, Gene applies a white base first.

2. Now the Aero-Lac silver base is applied over the white, covering in one coat, and is blended into the stock Ford silver on the front fenders.

3. The silver and gold bases used under candy paints have more particles than other metallic paints and need special care in stirring. See how much was on the bottom of our silver?

4. Winfield applied several different candy colors over the silver base in the flames, fogging and shading the edges with the darker colors.

5. Aero-Lac's Aero-Flake is mixed with clear acrylic or candy colors.

6. Flakes are usually applied at a low pressure (about 25 psi) to keep them from bouncing off or standing on end in the clear. An even spray is essential to most custom painting.

number of coats and of course is quicker to apply. Pearls also come in different colors, although they all look white in the jar. There is normal white pearl, and also gold, red, blue, and green pearls. The non-white colors are referred to as "cast" pearls. This has nothing to do with "casting pearls before swine" but it means that these colored pearls have one basic color, and a secondary, or cast color which is only seen in certain lights. For instance, gold pearl has a golden-white basic color, but viewed from some angles reflects a bluish cast, and so on with the other three colors. Most pearl concentrates are made for use with lacquer paints only, but in recent years, the custom paint manufacturers have come up with universal pearl concentrates that can be used in any paints, including enamels and two-part plastic paints.

Carl Green gave us the benefit of a pearl technique he calls "high and dry." He uses a lot of air pressure (70-75 psi) and thins the pearl/clear mixture out with four or five parts of thinner. This mixture he sprays from about two feet away, just dusting it on and trying not to follow a definite spray pattern. Three of these dry coats should be sufficient, but remember not to touch the surface, because the particles of pearl are barely clinging to the car, they're more like a coating of dust. A fast-drying thinner is used on these coats so that the material is almost dry by the time it hits the car. Putting it on this dry keeps the particles from laying down flat, making for a more dazzling pearl effect when it is cleared. Since the pearl particles are much, much smaller than flake particles, the problem of particles standing on end doesn't mean you have to apply lots of clear to cover them. You have to be careful in applying the clear coats though, because if you get this pearl-dust finish wet, it will spot. Your first clear coat has to be sprayed on medium dry. When this has flashed off, you can go to a medium-wet coat, then later to a full, wet coat.

"ERASABLE PEARLS"

A problem with many custom

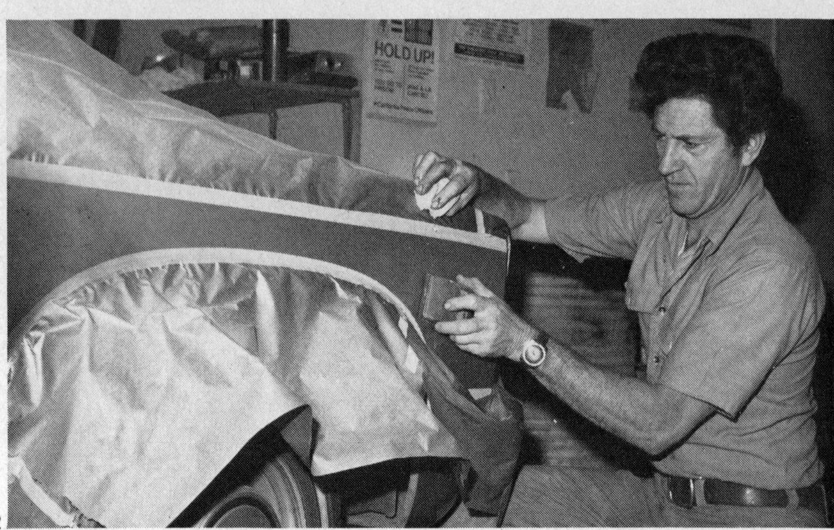

2

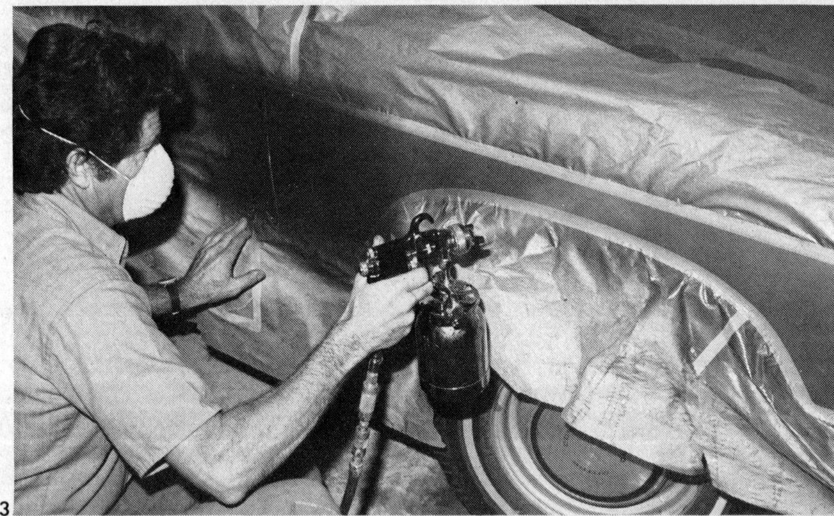
3

1

4

paints is that you can't decide after application that you don't like the effect, you're usually stuck with it. Fortunately for the custom van painters, the manufacturers have come up with tricky new pearls that can be erased and offer numerous special advantages. Metalflake Company calls theirs Eerie-Dess, while SEM Products calls theirs Design Colors. They are simply pearl pigments in a mineral spirit base, rather than a lacquer base. They can be used over lacquer, enamel, epoxies, or polyurethanes, and likewise, any of these finishes can be used over the erasable pearls with no ill effects. The mineral spirit base is very slow drying, allowing you to "work" the material after it has been applied. In a typical example, you spray a black or dark base color. Now the erasable pearl is either flicked on with a brush, daubed on with your fingers, or sprayed on (thinned 25% with mineral spirits, you cannot use lacquer thinner). It won't look much different from a standard pearl until you start to play with it. After spraying, you can pat it with aluminum foil or Saran Wrap which creates weird designs. Where the folds of the foil or plastic wrap have touched the surface, some of the erasable pearl has been moved aside to expose the dark background, leaving unusual "grain" or veining. You can also brush the erasable pearls on with a brush whose bristles have been deliberately cut to odd lengths. You can experiment to your heart's content, because you have time, and if you don't like how it's turning out, you simply wipe it off with mineral spirits and start again! The mineral spirits won't hurt your lacquer or enamel bases. The five basic pearl colors are available in this medium, but you should wait about a half hour between applications of different colors to the same panel. When satisfied with the results, let it dry for 24 hours and you can clear over it with any kind of clear, lacquer or two-part. The effect is unusual, to say the least, almost like marble (see the color section for an example), and has found fast favor with the van painters.

Like the custom painters they serve, the custom paint manufacturers are a creative lot. New products keep coming along and the painters keep finding new ways to use the standard custom paints. We've given you the basics of the four most common paint materials, and a photo sequence of a typical job involving some of these paints. If you've got the patience and fastidious working habits to get along in the custom painting field, you'll find creative work can be it's own reward, but also a lucrative, ever-expanding field for the professional spray-gun artist.

1. Before proceeding with any other work, use a blowgun to shake any loose flakes from your masking paper, or they may fall into a later coat.

2. Enough clear is applied over the flakes and candy so that final wetsanding prior to rubout doesn't get down to the irrepairable candy.

3. In our case, four coats of candy blue were applied over the flakes and three coats of clear after that.

4. Our "interior" work was masked off so that the outside border could be sanded, cleaned and shot with a few light coats of candy blue.

5. In cases where there is a heavy buildup of paint layers, the masking tape is easiest removed after spraying a highly-thinned coat of clear along the edges. The thinner will soften the paint enough so that it will tear off cleanly at the masking tape edges.

6. These tape edges on the flames are carefully and lightly sanded with a Scotchbrite pad or #400 paper to knock them down before clearing.

7. The entire job is covered with several more coats of clear acrylic lacquer for depth, gloss and protection. See page 170 for our results in color.

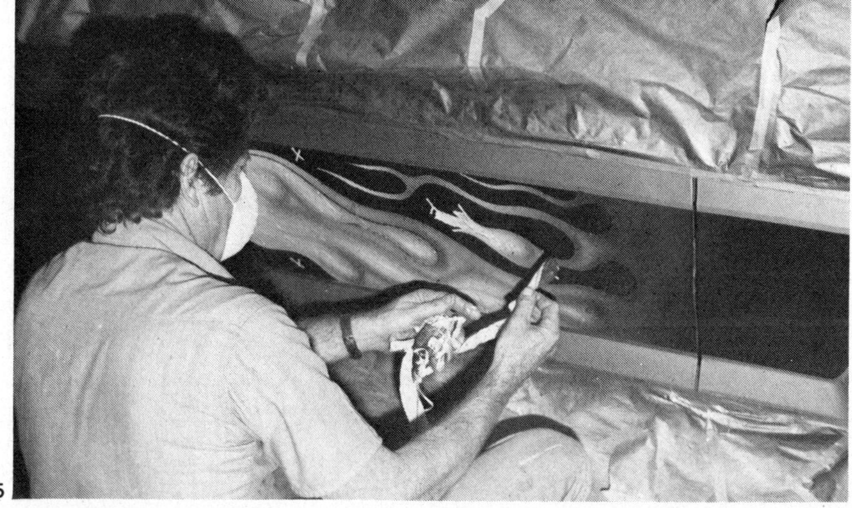
5

6

7

Pinstriping Tips

The art of fine-lining.

by Richard Busenkell

We mentioned the history of pinstriping in our earlier article on paint history, so here we'd just like to show some interesting effects achieved by famous customizing stripers, and pass on some of their techniques.

Let's start off with something surprising. One thing which all stripers seem to agree on is that you do not need a steady hand. Maybe they're a special breed, and don't realize how steady their hands are compared to the rest of us, but stripers insist that their steadiness is a result of proper bracing technique. Most use only one hand, using either the third finger alone or the third and little finger paired together to follow a car's character line in guiding the brush. Many use a two-handed grip, however, and if you turn out to be one of those, don't feel bad, for you're in good company—Von Dutch uses two hands. You simply have to find the grip that makes you feel most comfortable. And to do that means you have to practice—a lot.

Another point about which stripers are nearly unanimous is that the single most important thing to get right is consistency in brush loading and paint viscosity. You cannot stripe consistent lines if you load the brush differently each time or if the paint viscosity is not exactly right and exactly uniform.

Let's discuss some specific points that painters always ask about striping.

Brushes: The striping brush is called a sword or dagger. The sword brush has a round handle and the dagger has a flat handle. Two popular brands are Grumbacher and Mac. A normal paint supply house won't have them; try an art supply house or a sign painter's supply store. Sizes 00, 0, and 1 are best for striping.

Paint: A good oil-based enamel is best. Don't use lacquer unless you're going to stripe a lacquer-painted car, and even then enamel is better for striping. Lacquer dries fast, so if you must use it slow down its drying time with castor oil or a chemical retarder.

Paint Consistency: This is one of the key points about striping but it is hard to properly describe in print.

Set up a container of thinner next to your can of paint. For oil-base enamel use turpentine, and lacquer thinner for lacquer. If the enamel seems to be setting up too quickly and does not flow well, try a few drops of kerosene, but ONLY a few drops.

The paint must flow yet have a certain drag to give the brush its "rudder action." To achieve this—and consistent brush loading—dip the brush into the paint, then into the thinner. On a clean piece of paper or cardboard, work the brush, slapping and wiping it on both sides to distribute the paint evenly throughout the bristles. This must be repeated each time a line is started, even if the line is short. Experimentation and practice will tell you when the consistency is right; the flow will be even and smooth.

Brush Grip: It is easier to make straight pulls with a dagger and curves with a sword brush, for you have to be able to twirl the brush when making tight circles.

The normal grip is to hold the brush between thumb and forefinger, leaving the other fingers free to guide.

Practice: If you're a beginner, use a piece of glass as a practice surface; it's flat, hard, and can be cleaned and reused over and over.

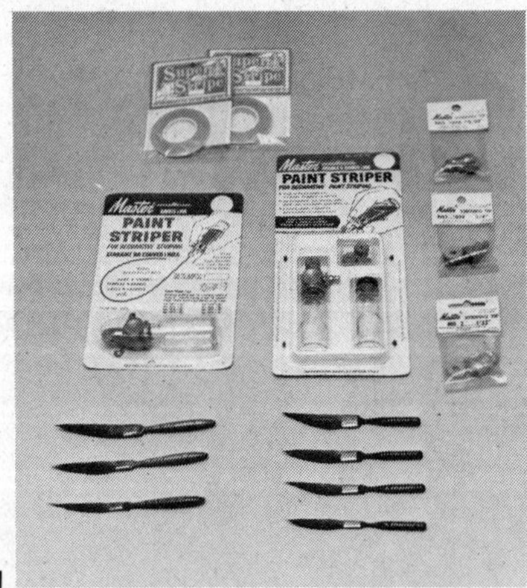

1

2

3

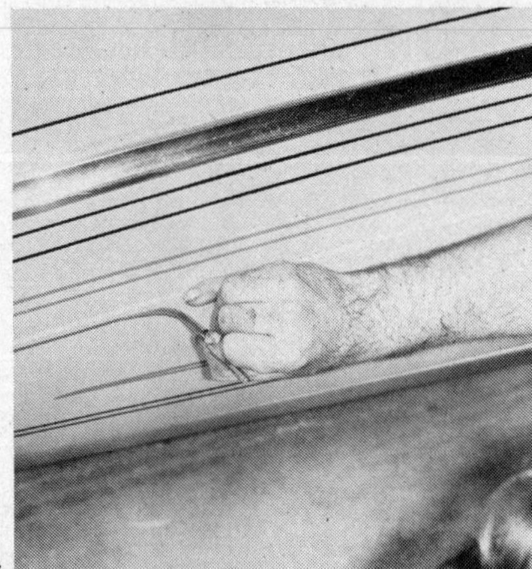
4

5

1. You can stripe many ways. For the faint of heart there's striping tape (top), then there are striping machines with interchangeable tips for lines of various width (center), while for the real artists there are various sizes of sword and dagger brushes.

2. Working the brush on a board to get exactly the right consistency is the mark of a real pro. Seelig's custom finishes sells striping paint.

3. Before striping the outlines of these custom paint shapes, the painter goes over the edges with rubbing compound. This cleans the area and reduces the height difference between the two layers of paint, allowing a smoother stripe.

4. Concentration and a guiding finger allow a striper to lay down a multitude of unerring straight lines. A strip of ¼-inch masking tape acts as a guide in the absence of a body line.

5. Veteran striper Andy Southard Jr. applies his talents to the nose of a Camaro. Striping accents car's basic body shape and character lines.

6. Pickup tailgates are excellent areas to display the pinstriper's art. Chevy Cameo Carrier (top) and two Ford Rancheros show diverse effects.

Cardboard or masonite should be used only if the pieces have a glossy coated surface, closely resembling a car's painted body. Don't use plastic; it has static electricity buildup which will foul up the striping.

Start by striping straight lines with the practice board laying flat. When you have some expertise, try curved lines and corners. Then put the practice board vertical—in a vise, perhaps—and try the same lines over again.

Thin straight lines are easier to control than thick ones. Pull straight lines as long as the practice board allows, striving for constant thickness. Always pull in one continuous stroke. Perhaps the hardest single thing to do is to pick up an interrupted line in the middle of a stroke; you'll find it's easier to erase the completed part of the line and start over.

Corners are the intersections of two straight lines. To get a sharp corner, don't stop the brush motion right at the corner; carry it past, then go back and erase the length of line past the corner.

Curves: Once you have consistency with straight lines, try curves. Start by trying to draw perfect circles with the brush. You'll find that with small-diameter circles it's rather easy to draw a good circular shape, but difficult to achieve a constant-thickness line, while with large-diameter circles you'll have the opposite problem. Try these circles in both clockwise and counterclockwise directions. If you find that you are markedly better in one direction than in the other, practice only in the good direction, and make all your curves that way.

In all curves, the brush must be rotated into the curve in such a manner that the flat side of the blade remains tangential to the curve. If you do not rotate the brush, you will find that the stripe rapidly becomes wider.

The consistency of the paint should be a little heavier for a curve than for a straight line. This is because the action of making the curve tends to open up the bristles of the brush, so a heavier consistency will hold the bristles to the same width as a straight line.

For a real challenge, try the letter "S". Get that down and all other curves will seem easier.

For any curve on a vertical sur-

6

face, start at the bottom and stripe upwards; this keeps the paint in the brush. If you were to start at the top and stripe downwards, the handle of the brush would then point downwards, and paint would flow out of it all over the handle and your fingers.

Line Tips: The central idea of striping should be forward motion. Striping ties together different parts of the car, making the vehicle more unified, but it should also impart a sense of purpose and direction. Senseless squiggles detract from the car's basic character, rather than enhance it. A formal, upright, conservative car with lots of angled body lines benefits greatly from long horizontal striping, as it lends the car a feeling of unity and motion, while a low zoomy car benefits from short striping which highlights the car's body and creates additional interest in its basically simple shape.

The usual way of ending a line is to let it taper to a point at either end, and this is the proper way to stripe an expensive luxury car. For a less formal or custom car, you can put a small arrowhead on the stripe's forward end or "feather" tips on the aft end, or both.

Sometimes it is necessary or interesting to break a continuous horizontal line. When you do, give the line a noticeable vertical break, and cant the break lines to the rear. This canting maintains the direction of the line.

The Car: The area to be striped must be spotlessly clean. This means removing all dirt, oil, grease, and wax with thinner or a special oil/wax remover available at automotive paint stores.

Make it easy on yourself with your first striping jobs. Pick a car which has definite character lines you can use to guide your hand, and make the stripes accent those lines. If you've done enough of that practice on a perfectly flat surface, you may be surprised at

1

2

3

4

how nice a stripe you can draw with a body line to guide you.

Think of obstacles sticking out of the car—door handles, mirrors, aerials, ornaments—as obstructions to fluid flow, which indeed they are. Instead of invisible air, imagine that fluid to be visible, like colored water, and then draw lines around the obstruction which would resemble the eddies left by it.

Part of the preparation of the car consists of picking a good place to stripe it. Remember you're using enamel, which takes time to dry. If the car is outside and there's any breeze blowing, you're going to have the same problem with dust and insects that you would have if the whole car had just been painted with enamel, only to a lesser degree. Stripe indoors if possible, and allow two days before washing and waxing over the new striping.

The striping is permanent and will last as long as the rest of the paint on the car, unless you get careless with compound or a buffing wheel. Remember that the stripe sticks up above the normal painted surface, and is therefore very vulnerable to damage from abrasives. The best rule is to allow absolutely nothing except pure wax anywhere near the striping. And pure wax means just that, not a combination of wax and cleaner.

Brush Maintenance: If your brushes are properly cared for, they will actually get better with age. When you are finished painting, wash your brush out completely with thinner. Then saturate it with normal automotive motor oil (motor oil is used because it does not decompose the natural materials of the brush and is readily available). Form the hairs into the blade shape. Lay it on a clean flat surface and stroke the brush flat so that the upper edge is straight and the lower edge is an even curve. Store it in a place where it will not be disturbed, and the brush will take on a "set."

Helmets and Gas Tanks: Motorcycle tanks and crash helmets are awkward objects to get a good grip on for striping. For helmets, sit on a chair or on the floor and put the helmet on your bended knee for support. For tanks, sit on the floor and clamp the tank between your legs.

That's it. Good luck with your striping. Don't forget: Practice is really important. It's also fun.

1. California's own Dean Jeffries shows what can be done to enhance the ordinary tail of a '56 Chevrolet.

2. Note how the blade of the brush is bent inward while following a curve. This necessitates a slightly heavier paint consistency than when striping a straight line.

3. Striping need not be confined to the car's exterior. Von Dutch brought these flames right into the cockpit, then added one of his strange little murals. Don't ask what it means.

4. This is a good position for pulling a long continuous line. Set your body to allow maximum arm reach.

5. If you don't feel comfortable with just one hand gripping the brush, try this two-hand grip. However, at least one finger must still guide.

6. Arrowheads and feathers can be used to start and end stripe lines. To stop the stroke, lift the brush gently up to taper the line.

7. A line interruption can still retain a sense of motion and unity if given some vertical connected lines canted toward the rear.

8. Practice straight lines by drawing parallel ones of different widths. Thick ones are harder to control in straightness and width than thin ones.

9. Practice making circles and S's to master curves. The constant rotation of the brush into the curve is the factor which controls the width.

10. Here are four variations for striping a corner. Once you establish a motif, stick with it; too many designs on a single vehicle are confusing.

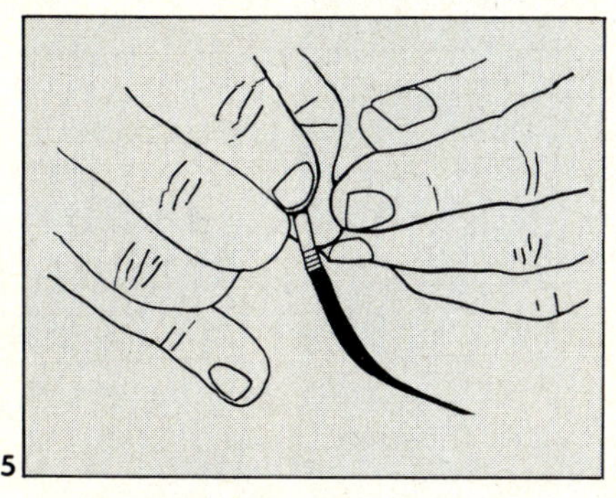
5

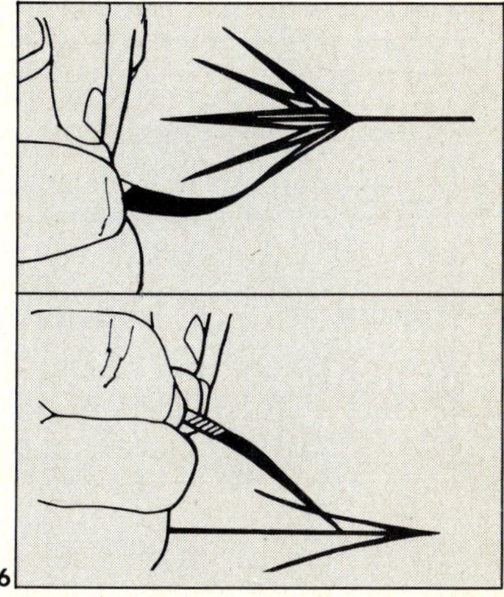
6

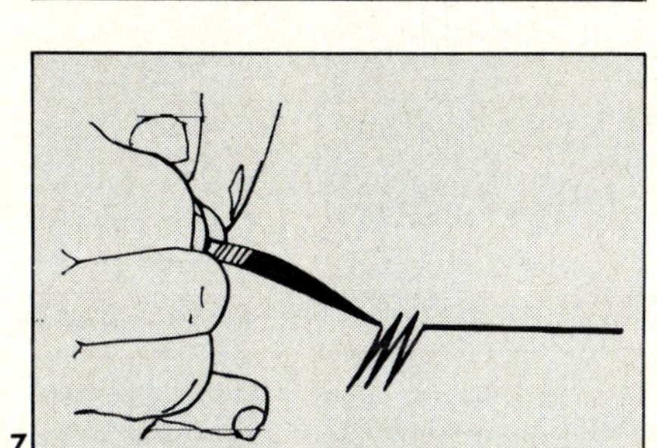
7

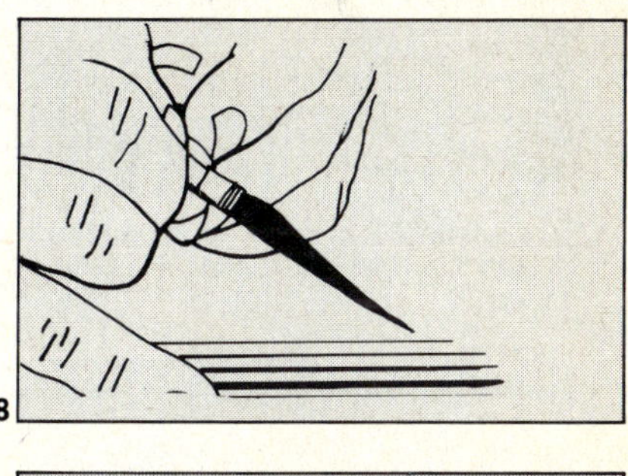
8

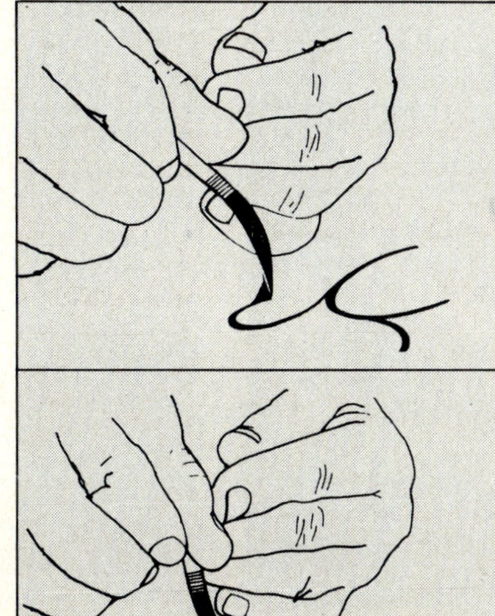
9

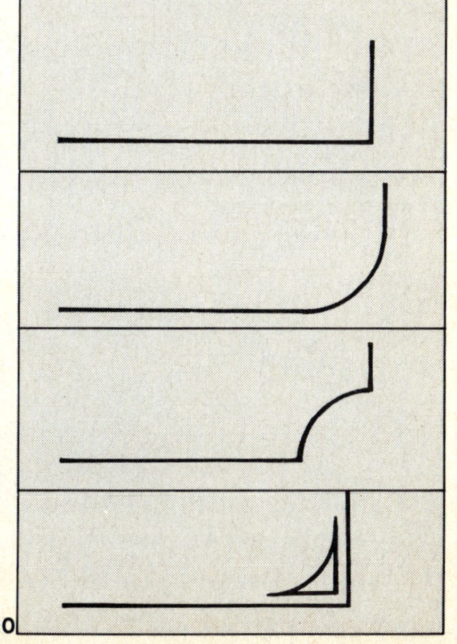
10

PHOTOS BY ERIC RICKMAN

Metalflake offers Beta tapes in three styles, TrimBrite has Prisma and Spectrum tapes, and SED Enterprises Space tapes are available in a rainbow of colors including gold and silver (chrome?). The generic name is diffraction tape and the tape's ability to reflect a rainbow of hues comes from the closely spaced grooves in the foil surface.

It has gained great prominence on the tailgates of pickups and the sides of custom vans. Its inability to conform to compound curves has severly curtailed its uses to club names, and marque identification. Although available pre-cut in the most popular manufacturer's names or individual letters, selection of pre-cut's invites membership in the banana club—and makes the user "one of the bunch." Individualize such logos and letters with shadow outlining. Better still, create one from scratch.

Tom Kelly, master sign painter and creative genius from The Crazy Painters, insists that marque logos are easy to stylize, and favors a bold script for designs in diffraction tape since the larger area takes advantage of the tape's best feature. When Kelly requested a palette to prove his point, Ted Jones Ford in Buena Park, California, responded with a handsome little Fiesta. Kelly branded the little critter on both flanks, added a few tasteful stripes to integrate the bold artwork and create a mildly personalized look, then handed it back to Ted Jones Ford in less than one hour.

His secret, aside from being able to visualize the finished logo before even beginning, is in not wasting time carefully trimming letters

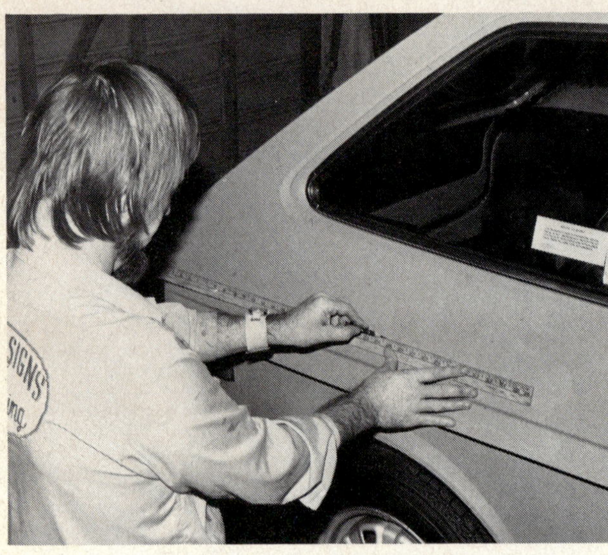

1 Deciding upon the size and location of the lettering, the surface is wiped clean with prep solution and a base line established using yardstick and Stabilo marking pencil.

2 Sheet of diffraction tape is affixed to table with masking tape and guide lines drawn to establish letter size. The Stabilo pencil is then used to sketch stylized logo on sheet.

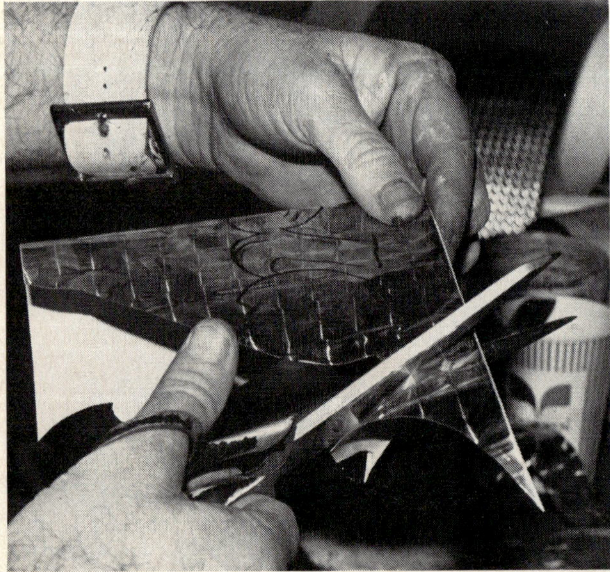

3 When proper proportions and letter spacing is achieved, the outline can be trimmed with a pair of scissors. Remember, it isn't necessary to cut out each individual letter.

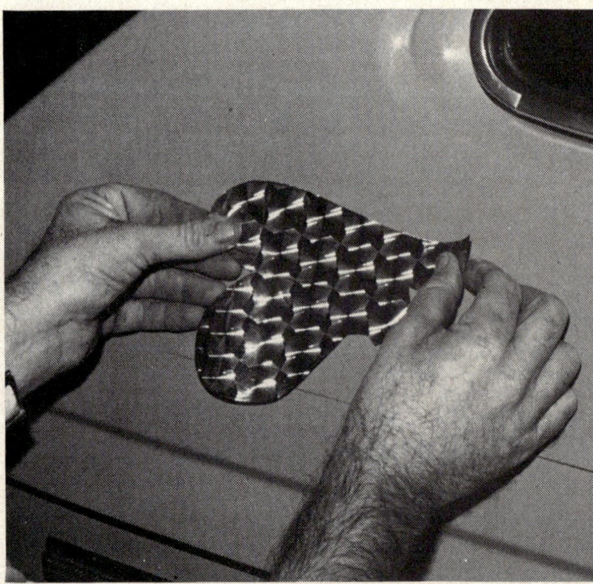

4 First letter is easily positioned on the baseline since the guide lines drawn in Step 2 are still quite visible. Simply peel off protective backing and affix it carefully.

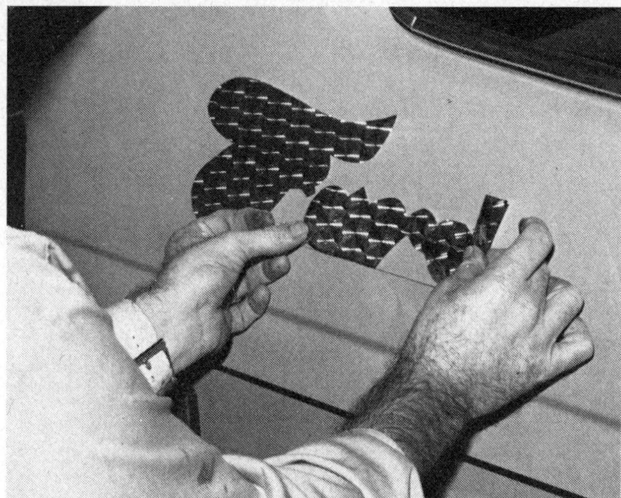

5 Ditto for the remainder of the lettering; which, in this case, is one not-too-well-defined piece. Take care to align lettering, because once adhesive touches paint, it's stuck!

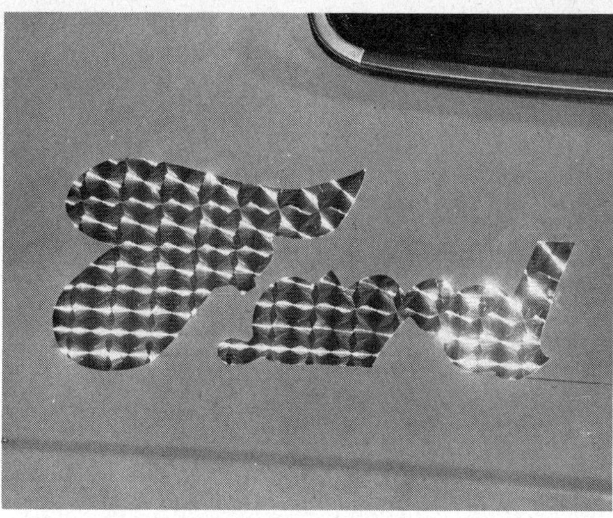

6 A soft rag, lightly dampened with thinner, is used to remove guide lines and prepare tape surface for paint. The logo looks very abstract now, but paint will give it form.

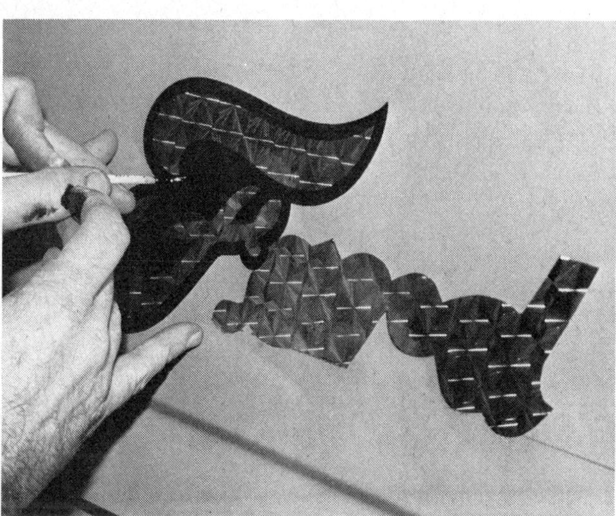

7 Paint is used to outline lettering and seal edge so that moisture can't creep under edges and lift tape. Mass of tape is filled in with paint and a letter 'F' takes shape.

8 After outlining and filling in, logo is given 3-dimensional appearance by thickening the outline. There are many styles that can be used for shadowing letters, too.

9 A few highlight streaks to the shadowing with white paint and the lettering really pops out. An airbrush permits precise control over this step in the proceedings.

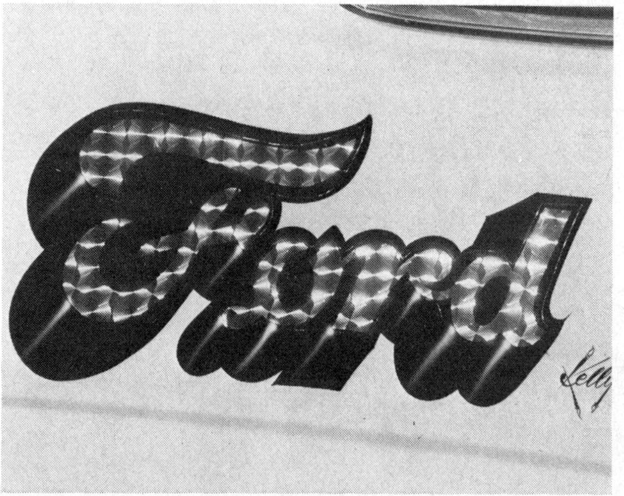

10 This angle shows how paint outline overlaps edges of diffraction tape. Coating the whole works with clear isn't necessary since diffraction tape is already coated.

HOW-TO: DUPLICATING WOOD

PHOTOS BY ERIC RICKMAN

Bob Eldridge has been using powders to simulate wood grain for more than 20 years. He can mix them to duplicate any type of wood and any grain pattern. He was taught this technique by his father, who many years ago worked in Detroit woodgraining dashboards and window moldings.

The materials needed are few and relatively inexpensive. The powdered pigments are available at most stores selling artist's supplies. Bob uses the products of the Permanent Pigment Company from Cincinnati, Ohio, but mentioned that the American Crayon Company (a division of the Joseph Dixon Crucible Co.) sells a fine product called Art Powder Tempra ($1.10 for 1 lb. can). The colors selected depend upon the wood to be simulated, but yellow ocre, cadmium orange, raw and burnt umber, raw and burnt sienna, Van Dyke brown, cocoa brown, carbon black, charcoal black, cadmium red and red ocre are among the most used. Brushes needed include standard and Badger Tail styles of good quality. For standard natural bristle brushes, a 5½-in., a 3½-in. and a 2-in. width will do nicely. Obtain the camel hair Badger Tail brushes in a 4-in. and a 2-in. width.

The sizing solution can be any mildly acidic liquid from vinegar to stale pale ale. Bob prefers to use white vinegar, but for the job illustrated on these pages, he used club soda. The only other items required are a box to contain the powders and an old coffee can to contain the sizing solution.

Looking at the step-by-step photos may give the impression that Bob Eldridge's method of wood graining is disarmingly simple to do. It is . . . for a master of this almost forgotten art. Bob suggests that anyone desiring to become proficient with this technique practice on an old fender—and practice, and practice, and practice!

The almost forgotten art of pinstriping enjoyed quite a renaissance when it was discovered by custom enthusiasts. The same thing might happen with this woodgraining method. Bob admits the field is not overcrowded and ample opportunities await in not only the customizing field, but in auto restoration shops as well. In fact, Bob was working on restorations for an automotive museum when he was lured away by Satyr Enterprises (van conversion/customizing specialists).

1 Wipe down the areas to be paneled with prep solution to remove wax, polish, etc., lay out design with ¼-in. wide masking tape. Take measurements and mark if needed.

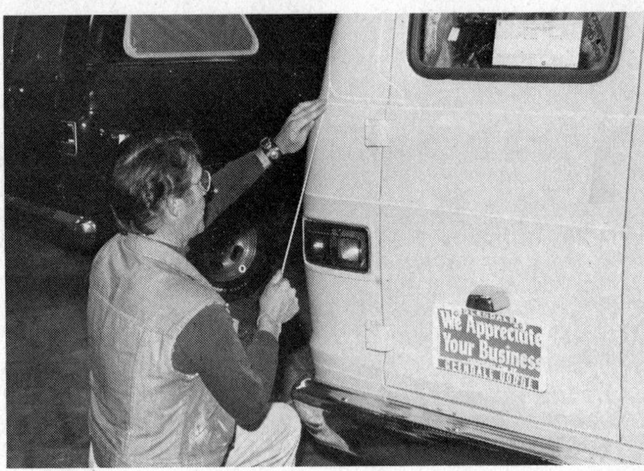

2 Adjust tape and recheck overall pattern until satisfied. Door handles and taillights must be masked off, but rear door hinge will be given wood-grain treatment.

3 Although powder technique doesn't create any overspray problem, vehicle is usually masked off for application of basecoat. This vehicle was tan, so no basecoat needed.

4 However, a black border was wanted to define the panels, so the masked border was fogged in. Blank area to right of painter has been reserved for bay window.

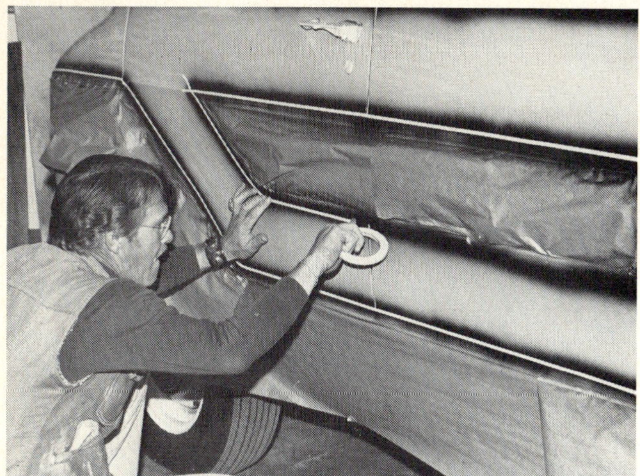

5 To achieve black border for panel outline, ¼-in. wide tape was laid on after black fog coat had thoroughly dried. Any color could have been used for the border.

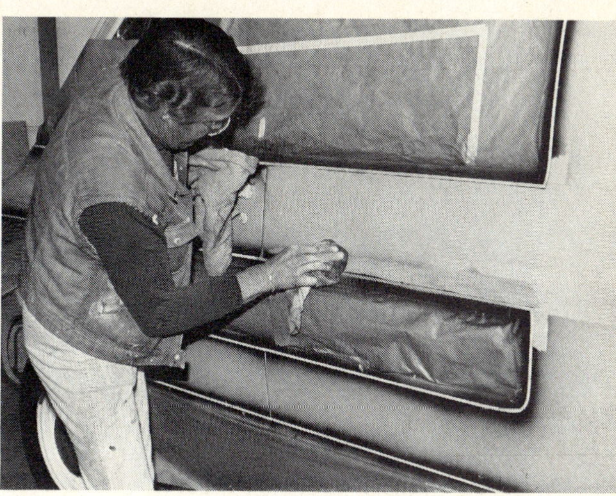

6 Once the ¼-in. wide masking tape has been placed and carefully burnished down, the overspray is removed with thinner and the paint roughened with a Scotch Brite pad.

7 Once the excess black is wiped off and the panels roughened, sizing can be applied. Work white vinegar or club soda into surface until it flows smoothly in a sheet.

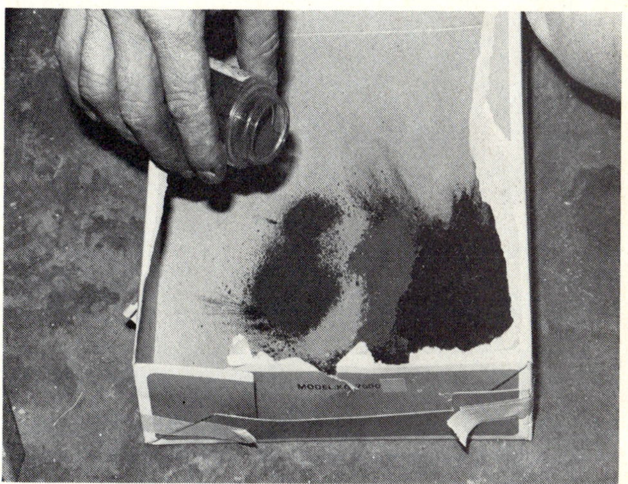

8 Any old cardboard box will suffice as a powder container and mixing palette. Colors used here (from right) charcoal black, raw sienna, Van Dyke brown, burnt umber, cocoa.

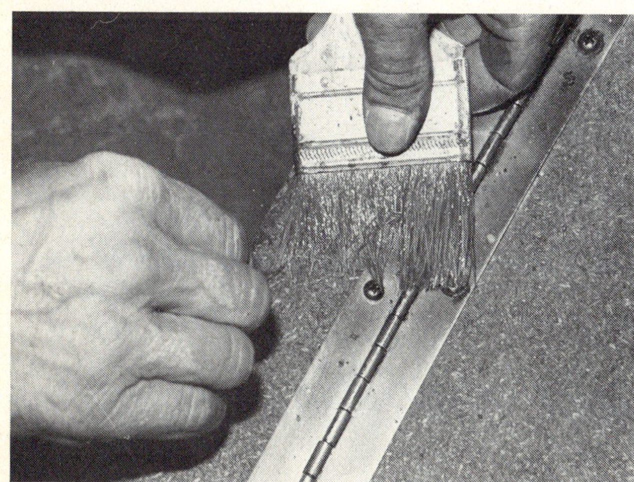

9 To obtain random streaks that will be refined into wood-grain pattern, brush is modified by hacking away uniform edge with a single-edge razor blade. Go easy at first.

10 Artfully(?) modified brush is now ideal for applying powder, but little else unfortunately. The wider the panel to be wood grained, the wider the basic brush is.

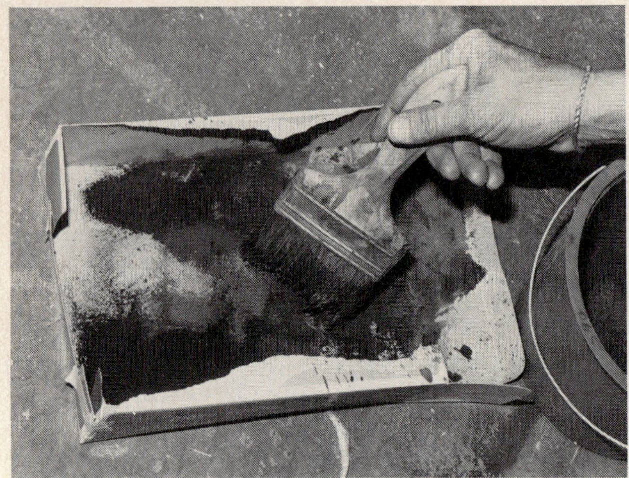

11 To begin, dip brush in vinegar or club soda sizing and touch to edge of powders simultaneously. Use far corner of box as palette to muddle the powders on the brush.

12 Loaded brush is pulled along the now dry body panel in one direction only. Remember, this is the start and it can be gone over and over until effect is pleasing.

13 The Badger Tail brush is used to stipple the waves, then the waves or lines are softened by cross-brushing. Color blending and brush handling are keys to this method.

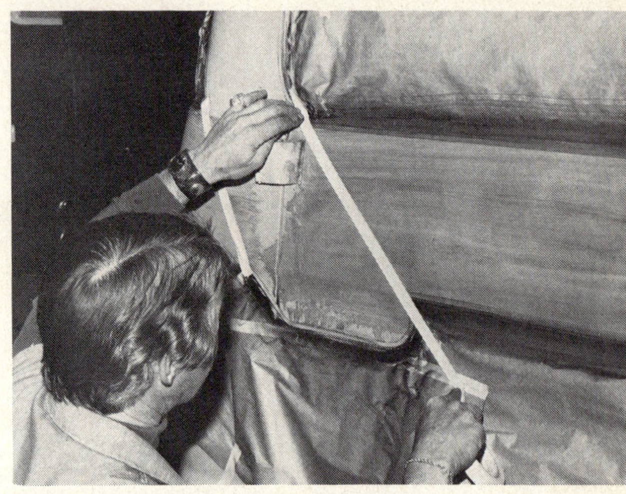

14 Even though the pattern can be removed with sizing at any time until the clear is sprayed over it, masking tape can be applied to the dry pattern without lifting it.

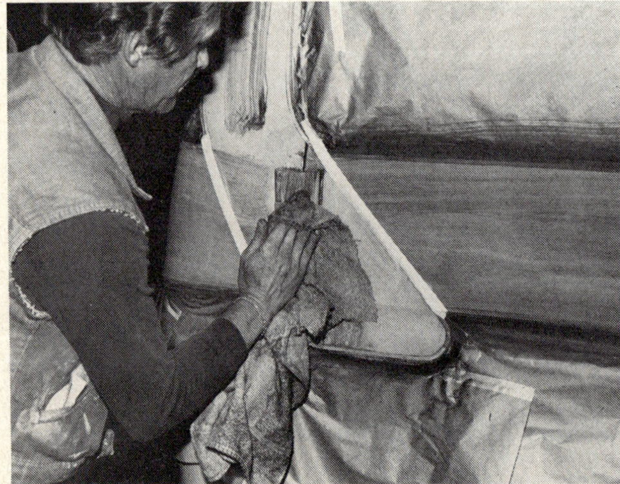

15 A panel of redwood is going to overlay the fruitwood panel seen to the right of the masking tape. Clean off new area with sizing and wash panel with sizing solution.

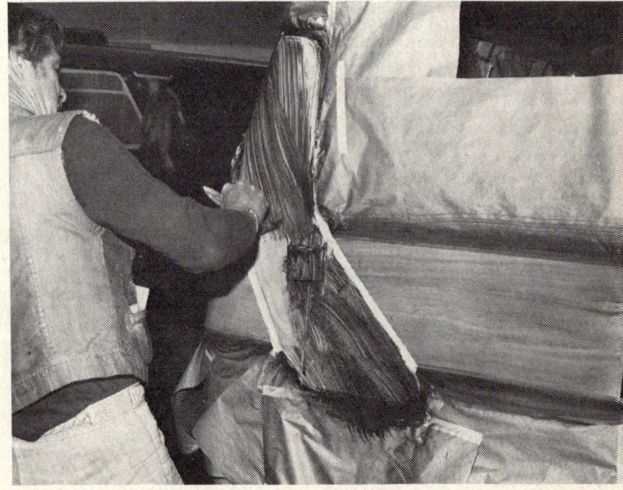

16 A different color mix and a different "painting" style create redwood overlay. Turning brush slightly edgeways will give closer grain. Experience comes with practice.

17 Again, Badger Tail brush is used to put finishing touch on the pattern. Hardware should be removed, wood-grained or made to look like they come through a flaw.

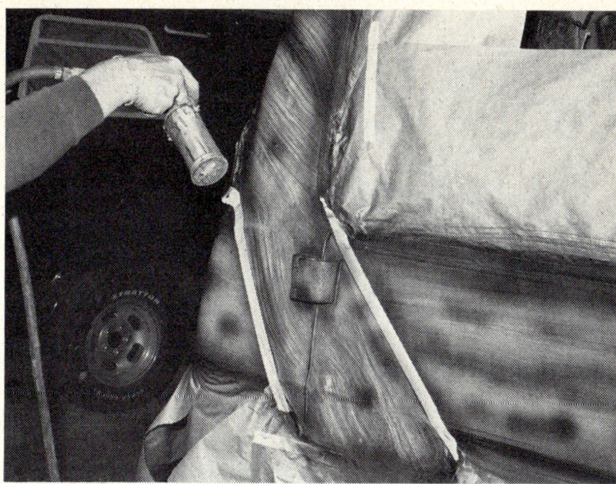

18 In the old days, master wood grainers sometimes used their chewing tobacco juices to color and tone the pattern. However, spray gun will highlight edges, etc.

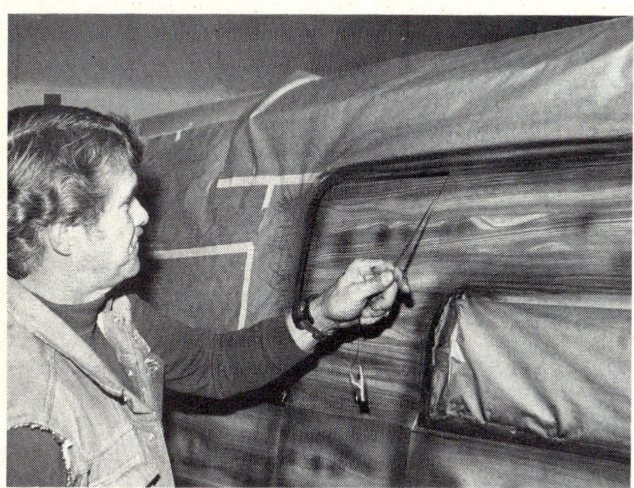

19 After highlighting with Van Dyke brown lacquer, ¼-in. masking tape is removed to expose black border surrounding the wood-grained panels. Transition is tough.

20 Application of about eight coats of clear acrylic lacquer—sanding lightly between each coat—really makes the wood-grain pattern jump out and shout.

21 With masking paper removed and the clear coat dry, the new generation woodie is ready for bay windows and other custom tricks that will make it a Satyr standout.

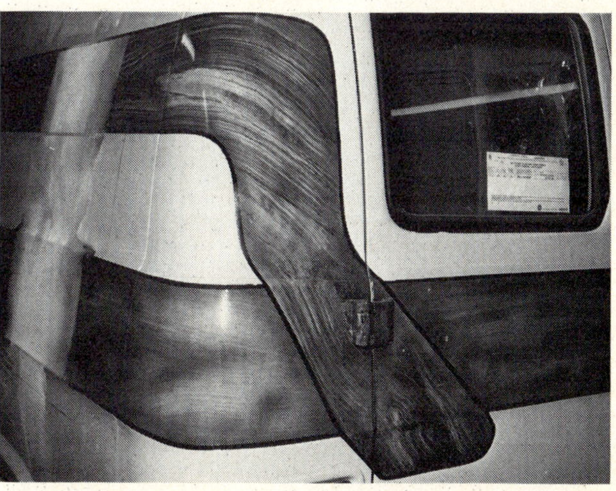

22 Cross-over pattern worked quite well, but black and white doesn't do justice to this realistic technique. For full effect, turn to color shot No. 8 on page 175.

HOW-TO: FREELANCE PHOENIX

PHOTOS BY ERIC RICKMAN

Murals aren't limited to desert scenes on the sides of vans or outrageous abstracts on the tailgates of pickups, because owners of other types of vehicles also enjoy the taste of individuality a custom painter can provide. One case in point was the owner of a Pontiac Firebird who wanted to adorn the hood of his car with the famous TransAm Firebird emblem (a Phoenix). The large decal can be purchased at the parts counter of any Pontiac dealership, but the big problem is properly applying it. Then, too, these decals tend to lose their shiny newness quickly in the polluted Los Angeles atmosphere.

The Firebird owner thought it over and decided to pay a visit to The Crazy Painters to discuss an alternative idea. The owner outlined his plan to Tom Kelly: A custom painted Phoenix similar, but not identical to the factory's decal. Something a little more muted in color but a bit more flamboyant in style was decided upon. Kelly agreed to freelance a Phoenix for the Firebird owner and pinstripe the car using the same colors as in the design to tie the whole thing together. This was an innovation the owner hadn't thought of, but quickly accepted.

Things then got right down to the nitty-gritty: Price! Tom Kelly, being an experienced pro, was able to quote a price for the whole job that was just a little more than for the cost of the factory decal alone. The owner was delighted and an appointment was set up. Here's how it went from there.

1 Hood was wiped down with prep solution, then a Stabilo Pencil was used to draw general outline of Phoenix. Masking follows outline; paper guards against overspray.

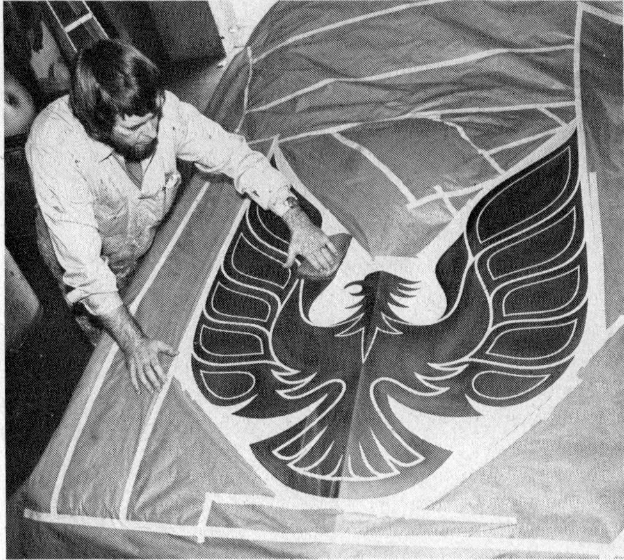

2 All masking tape was then burnished down with a roller, and exposed surface scuffed up with a 3M ScotchBrite pad. Design is similar but not identical to Pontiac decal.

3 A Binks #15 spray gun was used to apply blue basecoat that is nearly white. Progressively darker shades of blue will be sprayed over this base. Car is metallic blue.

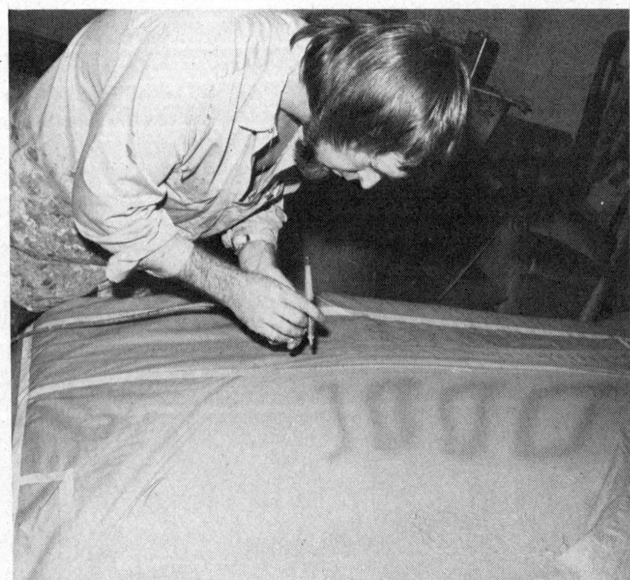

4 A Paasche Model VL airbrush is used to spray next darker shade of blue. Painter must be able to visualize final product before even beginning. Yes, it requires skill!

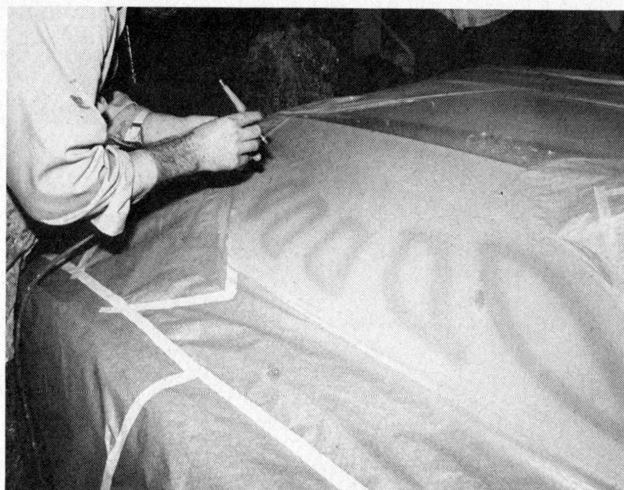

5 The VL is a double-action airbrush which allows control of both air volume and paint volume with single button. Painter can release small amount of paint for target check.

6 Broad fan or narrow streaks require nothing more than a sensitive trigger finger with double-action airbrush. Still darker shade of blue is fogged along inner edges of design.

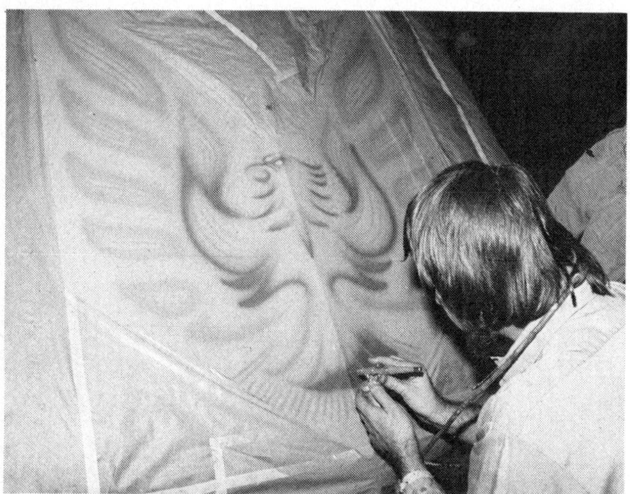

7 Air hose to airbrush is draped around neck to avoid dragging it across freshly painted design, and painter works from the center of design toward the outer edges.

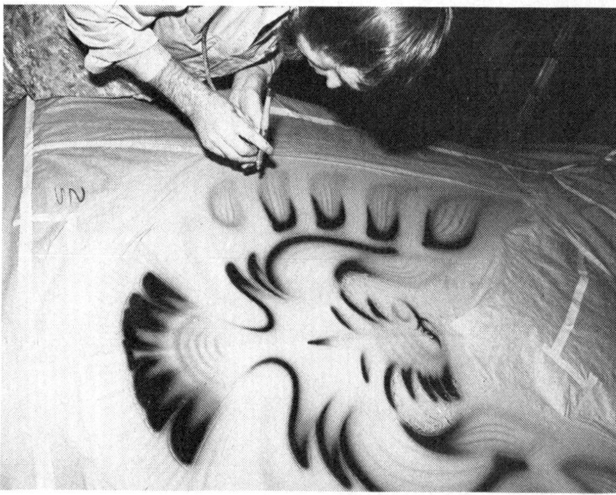

8 The final shade (a deep blue-black) was applied next and confined to inner and lower portions of the design. Note the freehand work that delineates the eye of the Phoenix.

9 After the masking tape was removed and the paint given time to dry, the design was outlined with pinstriping and white highlights added to the bird's eye and crown.

10 Symmetrical positioning was achieved by eyeball alone. Contrast of design with color of car is greater in black and white. To get full effect, see photo 4 on page 176.

COBWEBBING AND VEILING

PHOTOS BY ERIC RICKMAN

Like flame paint jobs, the popularity of cobwebbing and veiling comes and goes. Both techniques lend themselves well to use on fairly large areas or to adding interest to rather wide borders. Cobwebbing and veiling works well when used over Metalflakes and pearls and can be dramatic if covered with light shades of candy color.

Cobwebbing is accomplished by spraying unthinned acrylic at low pressure (15 to 20 psi). Lack of pressure and dilution of the paint results in the acrylic being spit out of the gun in stringy globs which resemble cobwebs laying on the surface. Needless to say, dark cobwebs on a light base color provide the best effect. The type of spray gun isn't as critical as the need to use a large nozzle/air cap combination. The webbing should be applied onto a wet clear coat (add thinner or even retarder if necessary) and allowed to thoroughly dry before attempting to spray with clear or candy. The finish must be built up with candy and/or clear as with a Metalflake finish.

Veiling requires the use of a special veiling gun and veiling paint, in addition to higher air pressure (18 to 30 psi). The syrupy veiling lacquer is mixed with color toner and is shot onto the surface in virtually a continuous stream, so constant and precise movement is important. It is wise to practice on scraps of paper or cardboard until air pressure, distance from the subject and correct arm (not wrist) movement yield the desired pattern.

Our photos were shot at Walt's Studio of Style in Van Nuys, California; long considered tops in practicing these techniques.

1 The panel to be cobwebbed or veiled is given a base coat of opaque, Metalflake or pearl, allowed to dry then sprayed with several coats of clear to make erasing easier.

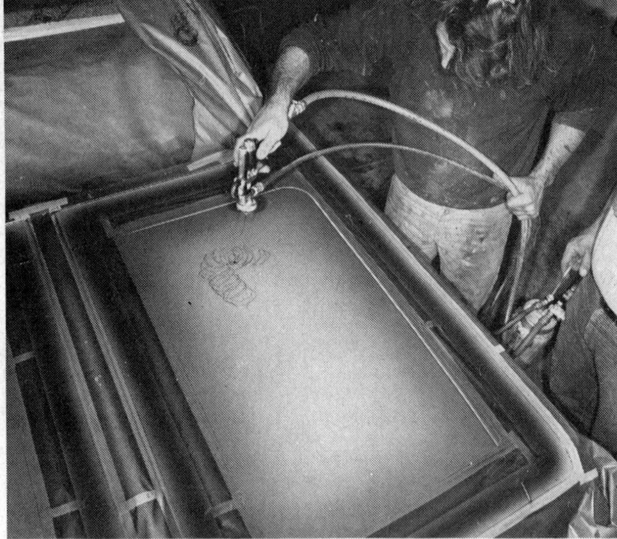

2 Veiling can be applied to dry surface, but applying to wet surface (see text) avoids smearing when oversprayed with clear. Note veiling gun and remote paint pot with dual hoses.

3 Veiling pattern is continuous line and thickness of paint can be varied from thread-fine to rope-thick. Cobwebbing is more a chopped fiberglass thread pattern with fine lines.

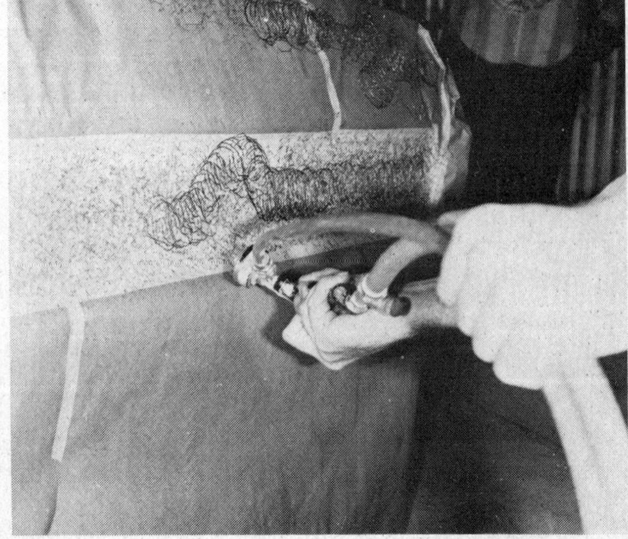

4 Yes, veiling and cobwebbing (seen beneath veiling) are easily applied to vertical surfaces, but control of pattern is extremely important. Experimentation and practice are musts.

PHOTOS BY ERIC RICKMAN

The creation of a mural on the side of a van or on any panel of any type of vehicle, especially one done freehand and using imagination as a guide rather than the copying of an existing piece of artwork, is not done without particular qualifications. First, the originator must have artistic ability—simply anyone cannot do it. Second, he must have a color sense, for while an experienced sketcher or line-artist may be able to lay down a decent drawing, he may not necessarily be able to subtly blend the proper colors together. Third, he must know the compatibility of paints and be experienced with both standard-type spray guns as well as finer air-brushes. Still another needed quality is patience. In short, vehicle mural creation is not for the novice.

The mural created here especially for this book was carried out by Tom Kelly, and the Ford Econoline 150 van on which the exercise was accomplished was donated through the courtesy of Ted Jones Ford, Buena Park, California. The van was as yet unsold and stood stark in its all-black factory paint job. A scenic landscape might have seemed out of place with an ebony-hued backdrop, so Kelly created a background of far larger proportions than the final scene, in order for the mural to have a "reason" for being. The background is framed with stripes and bands, all laid out in tape and masking paper as seen in the first few illustrations.

Tom used warm earth colors for his painting; browns, reds, oranges, to keep the tones in the same family. Before he applied tape or paint, though, he used a degreasing solvent to clean the area to be painted, wiped it dry, then lightly scuffed the surface with a Scotch-Brite pad, finally wiping it clean with a tack rag after blowing away lint and dust with compressed air.

Tom works—using only his own mind's-eye as guidance—using the darker colors first, blending them outward to the lighter ones. Since Tom is a pro, let's watch him work.

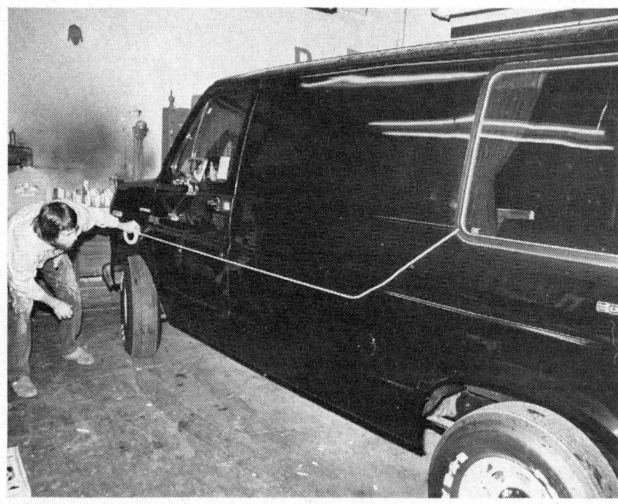

1 Preparation begins with the taping off of the central panel upon which the mural will later be painted. Narrow, ¼-in. tape is laid in a single strip parallel to a body crease.

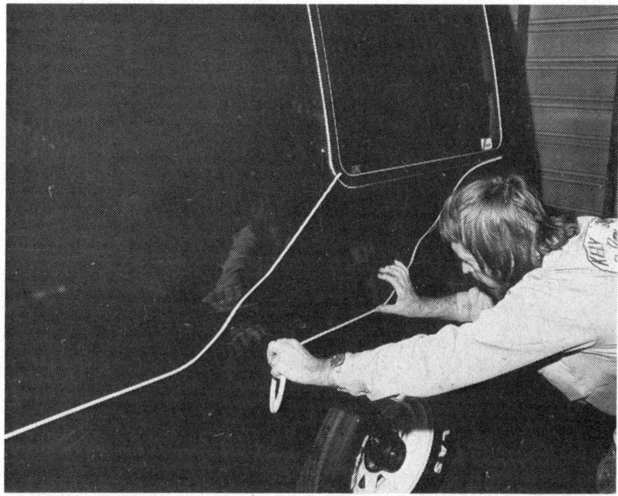

2 Another length of tape is pulled across the area; tape is unrolled and kept taught well ahead of the contact area, here formed by the thumb of the right hand.

3 Only the mural painter knows which taped-off areas will be protected by masking paper and thus be left black when the job is done. Kelly works without layout sketches.

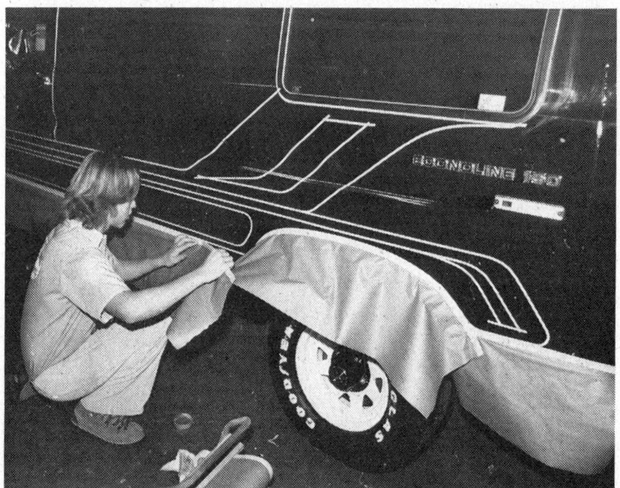

4 Masking paper, pulled from a roll dispenser which also applies tape along one edge, is laid along the ¼-in. tape. Paper strips as long as possible are used to ease unmasking.

5 With all outlining tape and paper in place, all areas to be painted are degreased, then lightly scuffed with a ScotchBrite pad, later air-blown then wiped with a tack rag.

6 Unlike art instructors' teachings, Kelly lays his darker colors first. His preference is what art shops term one-shot bulletin enamel (available in many brand names), thinned 60/40.

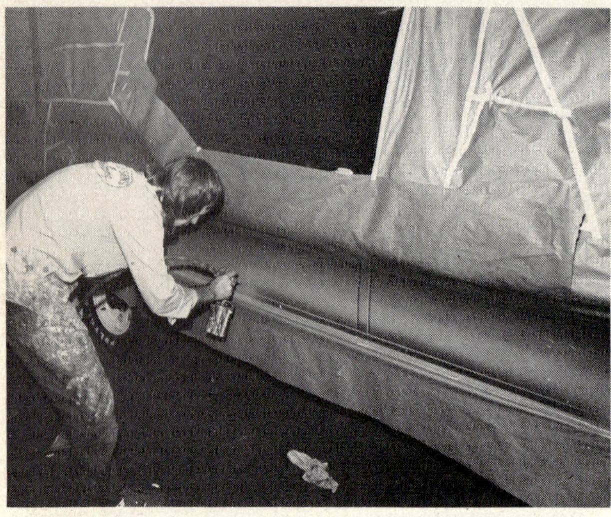

7 The larger areas are sprayed using a standard Binks #7 gun. Bands are fogged all over in a dark yellow, then edged in a dark orange. Black stripes are marked by tape.

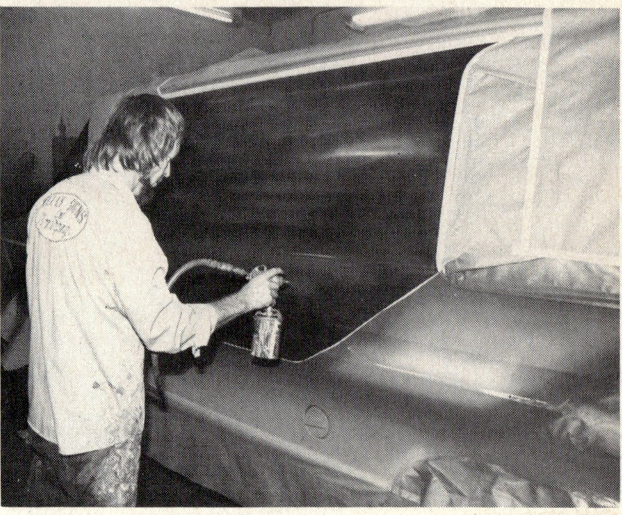

8 Protective masking paper has now been pulled off area to contain the mural, and here again Kelly lays on the background of his darkest hues.

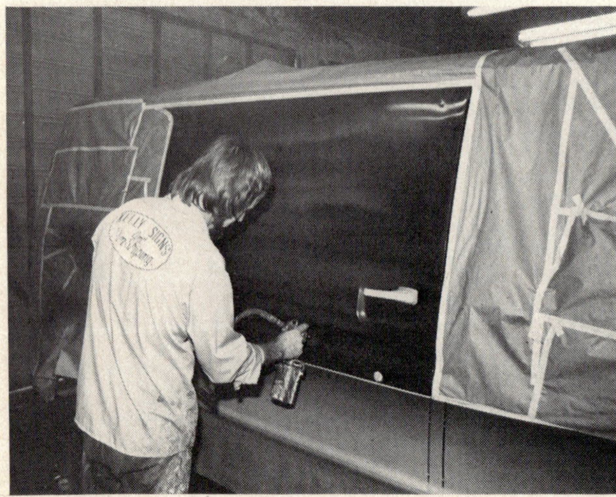

9 Barely discernable gradations of color will come to life later; for now Kelly is finishing up his swirly effect with the Binks gun at the base of one panel before doing the other.

10 Finer work is next, and Kelly readies the Paasche VL airbrush, mixing colors to his taste; making enough for all area of both sides in which the shade is duplicated.

11 The light, foggy areas Kelly did earlier now come to life as banks of clouds as he applies highlights with low pressure and a narrow spray, building up a 3-D effect.

12 The background now begins to emerge in hues of orange, red and brown and we can think we're seeing the development of a sunset—or sunrise.

13 Now the pastoral foreground scene is zeroed in on, and we can see the beginnings of a desert landscape as hills and crags emerge from the Paasche.

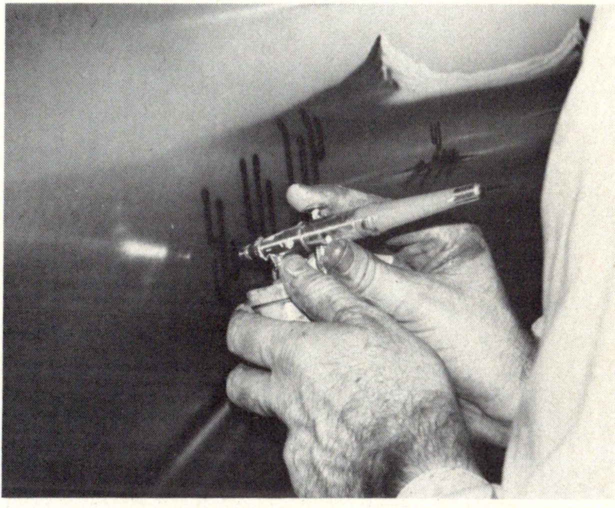

14 What's a desert without cacti? Ghostly plants are laid on in freehand, the artist building them in hushed tones where only he can foresee them.

15 As the final concept emerges, the painter can judge where and how to add interest touches; an eerie peak here, another cactus there, more highlights or shadows.

16 Final step is to protect mural with several clear coats. With no masks to cut, etc., job took less than 3 hours. It is shown in full color on page 176 (picture 1).

PHOTOS BY ERIC RICKMAN

The previous how-to on duplicating wood illustrates the realistic effects this method can achieve. In fact, it is very difficult to imagine it as not being a real wood overlay. But what if the owner wanted the "Gee Whiz—it's not wood but it sure looks good"—effect? Something, perhaps, a bit bolder and more moddish.

This can be accomplished easily with ordinary lacquers and an extraordinary technique that borrows heavily from the antiquing craze enjoyed by do-it-yourself furniture refinishers. The technique is that of applying a slow-drying dark color glaze over a light colored base, then wiping or (in this case) brushing the glaze before it dries to simulate the grain pattern desired. The moddish aspect is achieved by using a candy color base coat, and the wilder the base color, the more mod the finished wood grain will look. For example, a candy bronze base with a dark brown glaze would have a rich wood look, but candy red with a black glaze would be wild.

Steve Young at The Crazy Painters demonstrated another combination for us on a Chevrolet Luv pickup with a bed by California Step Side. Fortunately, the basic color of the pickup was yellow, so selecting the colors for the mod wood graining and the Saran Wrap mural detailed in the following how-to chapter, was simplified. And according to Steve, the most important aspect of custom paneling is in selecting the colors and designing the layout of the panels. His first piece of advice is to make the design flow with the style features of the vehicle. His second piece of advice is to practice this painting technique on an old fender hood or any other available body panel.

1 First of all, clean the surface right down to the paint with a good prep solution, make a few measurements and start outlining the paneling with ¼-in. wide masking tape.

2 Tape and cover the perimeter of the proposed paneling, then use a 3M ScotchBrite pad to give surface a bit of tooth. Get right against the edges of the taped outline.

3 A ground coat of light gray primer was shot first to seal base color of truck so it wouldn't come through candy. Also to afford a neutral base for the paneling to come.

4 The primer coat is lightly sanded with #400 grit wet-or-dry and then wiped down preparatory to first color coat. If flaws are found, spray again with primer and sand.

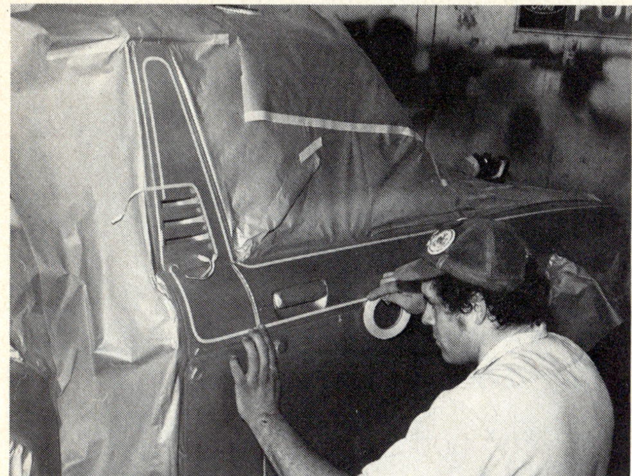

5 A heavy metallic-silver basecoat was sprayed next and allowed to dry thoroughly before bordering the panel to be grained with ¼-in. tape placed ¼-in. in from masked edges.

6 The basic area for the wood grain panel is now established. When the ¼-in. tape is finally lifted off, it will reveal a ¼-in. wide metallic-silver stripe around the design.

7 If some is good and more is better, an additional stripe will be just enough. Hence, ¼-in. in from the first stripe, another stripe is added—this one only ⅛-in. in width.

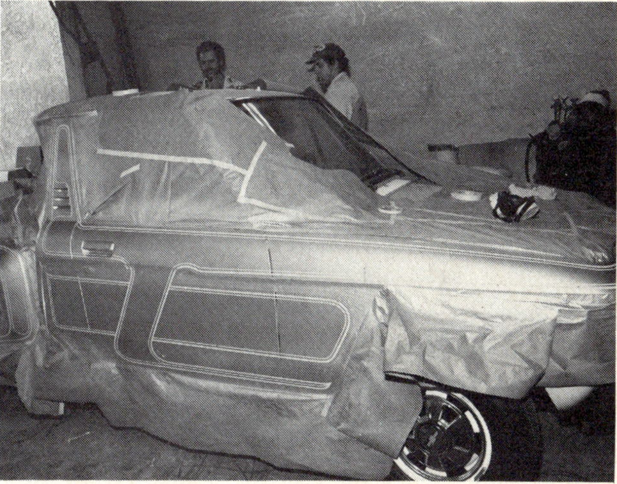

8 It looks complicated, and it is! Interrupted panel below the beltline is reserved for a how-to chapter following right on the heels of this one and must be masked off now.

9 Close supervision of this most important phase is an absolute necessity or the article following this one could be ruined. Know why they're called The Crazy Painters?

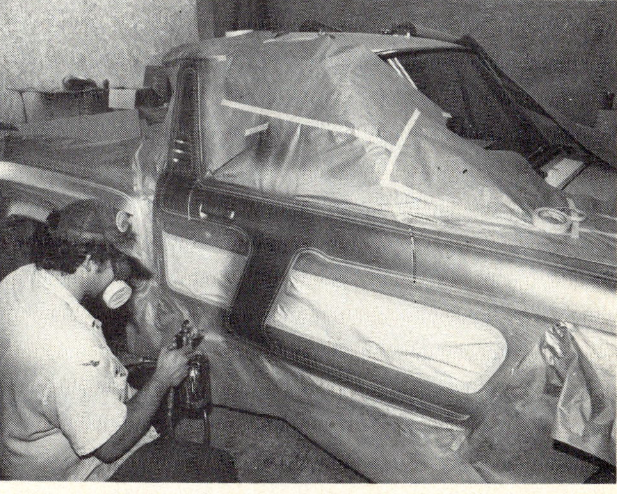

10 Since this is to be a mod wood grain, a candy orange is shot over the silver base. A Binks #7 spray gun and good quality respirator are used. Note even coverage!

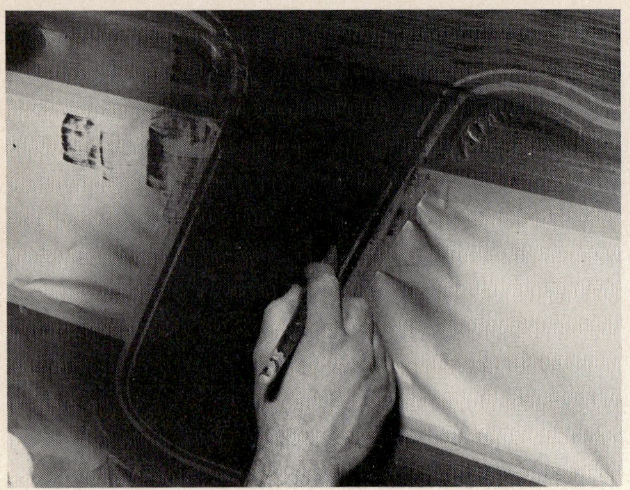

11 Black acrylic lacquer thinned with straight retarder is hand-brushed over the thoroughly dry candy orange base. Controlling brush determines wood grain pattern.

12 The retarder slows down the drying time and allows the pattern to be worked. Strokes should be horizontal and angling the brush closes up the grain pattern.

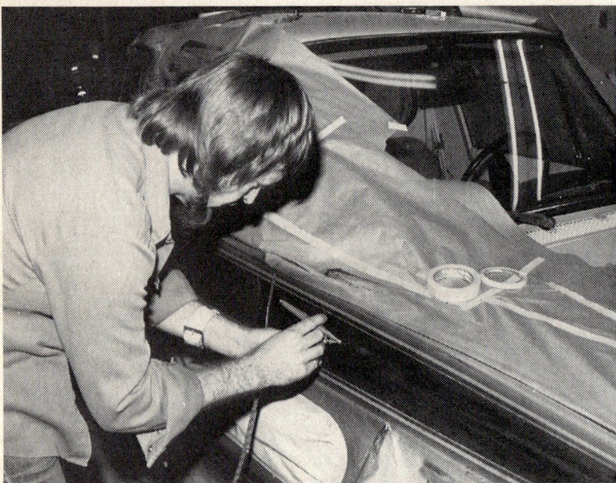

13 Hero supervisor (Tom Kelly) returns with Paasche Model VL airbrush and punches up the wood grain effect by adding knotholes, streaks and other details for more realism.

14 Once the glaze coat has thoroughly dried—a long time thanks to retarder—all but perimeter masking tape is removed and pattern sealed with Ditzler Delclear.

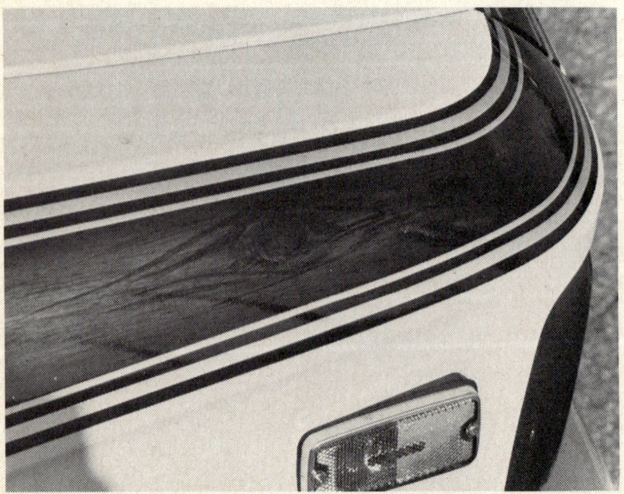

15 This job isn't done yet because there's one more panel technique to apply, but here is a sneak preview of what it looks like in finished form. Note interest added by airbrush.

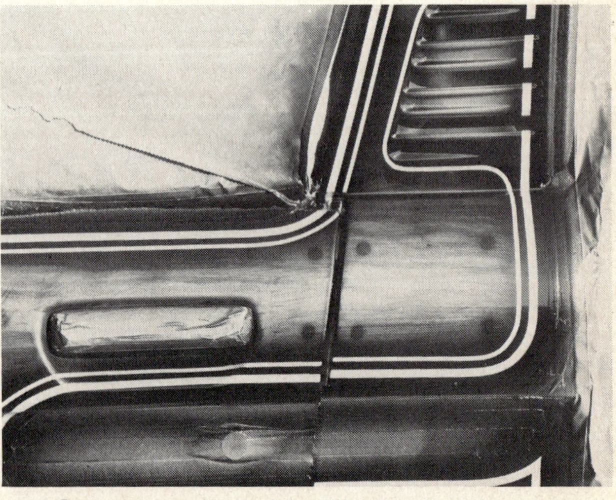

16 Pegs are easy to simulate with airbrush and cardboard template with hole in it. For the full visual effect, see the color picture (No. 4) on page 175. Next is a mod mural.

HOW-TO: SARAN WRAP MURAL

PHOTOS BY ERIC RICKMAN

Much can be done with this technique of texturing a slow-drying glaze coat. The wood graining shown in the previous how-to was merely one of many possible techniques for creating unusual patterns or designs. Then too, the effects illustrated here are accomplished with acrylic lacquer, not with unusual or exotic paint materials. There are, on the market, some very interesting new products that may cause custom painting to turn the corner and head in a new direction.

One of these exciting new products is manufactured by Metalflake. It's called Eerie-Dess. The other is made by SEM Products and is marketed as Aero-Lac Design Color. Both products are pearlescent acrylic paints with a mineral spirit base instead of a lacquer base. This retards drying and when sprayed, spattered or brushed over an acrylic lacquer base coat it can be worked in many ways. Of course if the resulting design is less than pleasing, a rag dampened with mineral spirits wipes the canvas clean and the artist can begin anew. There is one problem, however. The pearl pigment must catch the light or the incredible beauty of the color(s) remains hidden. Aero-Lac's Design Colors were textured with the plastic wrap method illustrated and can be seen in color on page 170, photo 4.

Although acrylic lacquers don't have the punch of pearl paint, lighting requirements are less difficult. So Steve Young of The Crazy Painters continues his paneling exercise on the Luv pickup borrowed from Bill Barnett Chevrolet in Compton, California. Steve is going to use Ditzler acrylic lacquer and Saran Wrap to create an avant garde mural.

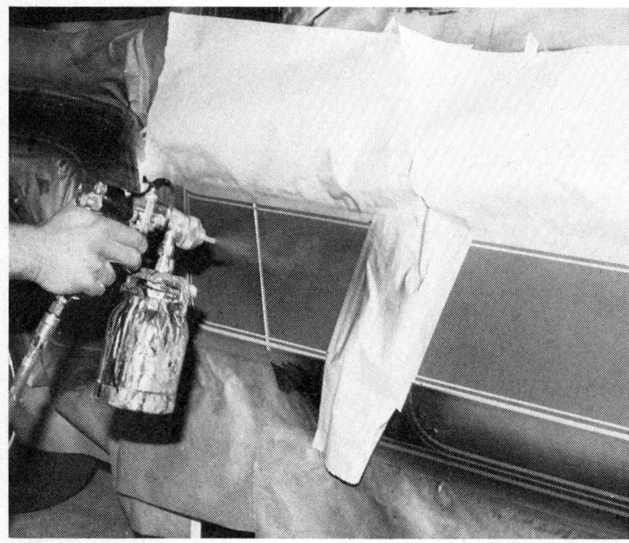

1 Picking up where we left off, the wood-grained panels are masked off—roll paper so it doesn't contact surface—and spray the panel with candy basecoat. In this case, blue.

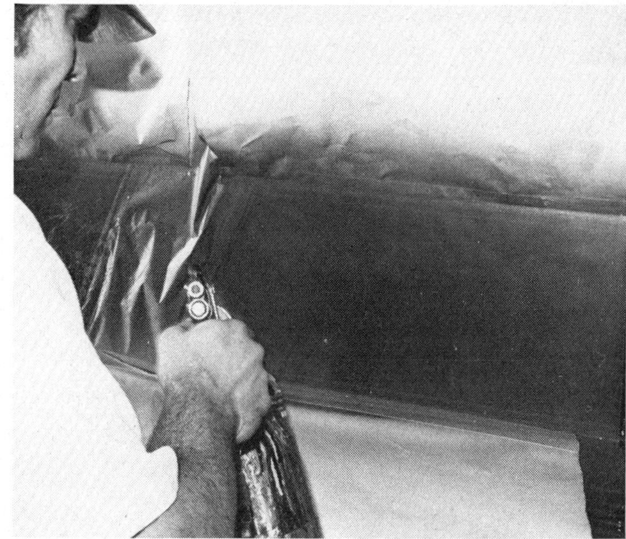

2 After basecoat has dried thoroughly, a light coat of black thinned with straight retarder is applied to the surface. Retarder will considerably slow drying time.

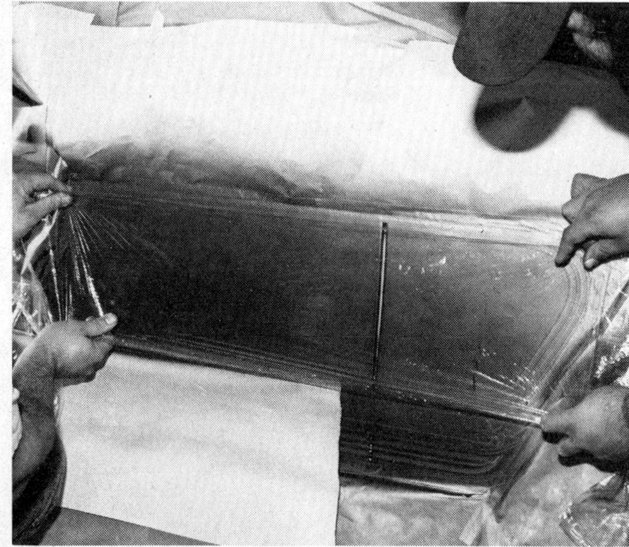

3 However, quick work is still a must—as is an assistant. Pull off a long strip of Saran Wrap and lay it against the black glaze coat which is still wet thanks to the retarder.

4 Use hands, feet or whatever to pat the plastic wrap against the wet glaze coat. The fingers can actually be used to paint through the plastic wrap.

5 Lifting the plastic food wrap away removes some of the black glaze and even some of the candy blue that's been softened by the retarder. There will be four colors.

6 Some areas where the plastic wrap didn't touch at all will be blue-black, others will mix and be dark blue, still others (where glaze has lifted) will be candy blue.

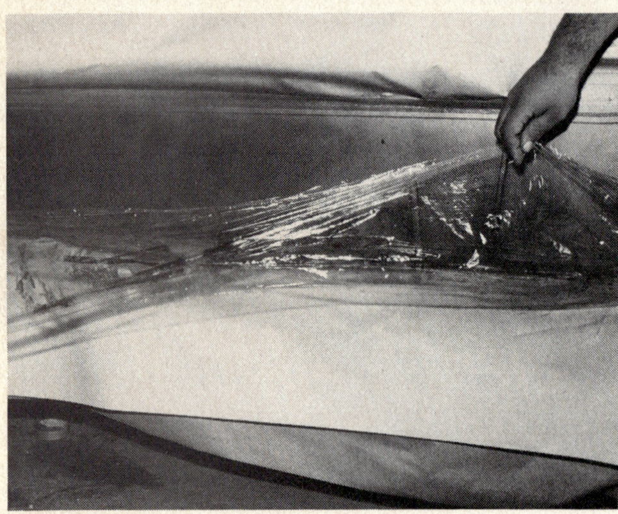

7 The panel on the other side of the vehicle is sprayed with the glaze next and the same piece of plastic wrap along with the same assistant are "pressed" into service.

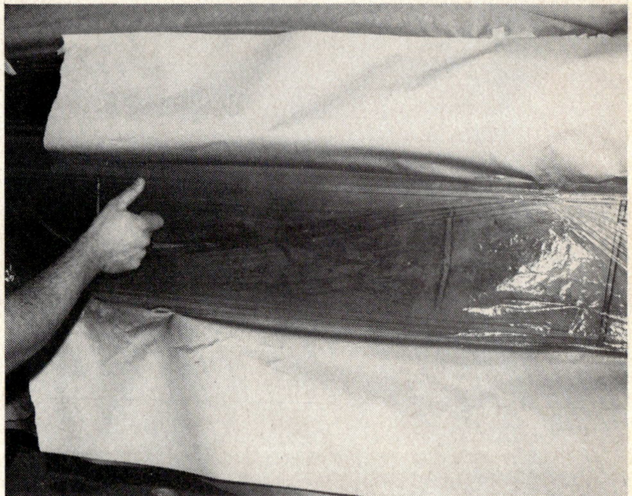

8 The more wrinkles in the plastic wrap, the more interesting the resulting pattern will be. Just make certain that the glazed panel is covered completely from edge to edge.

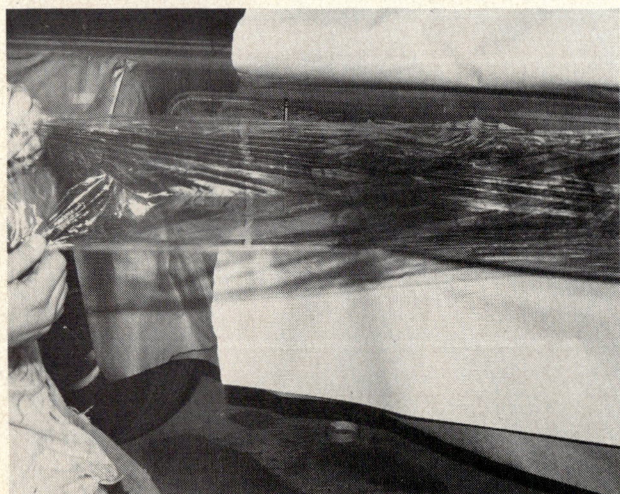

9 After patting, pounding or playing other creative games, the plastic is pulled away from the panel and another masterpiece is created. Remember, it can be removed.

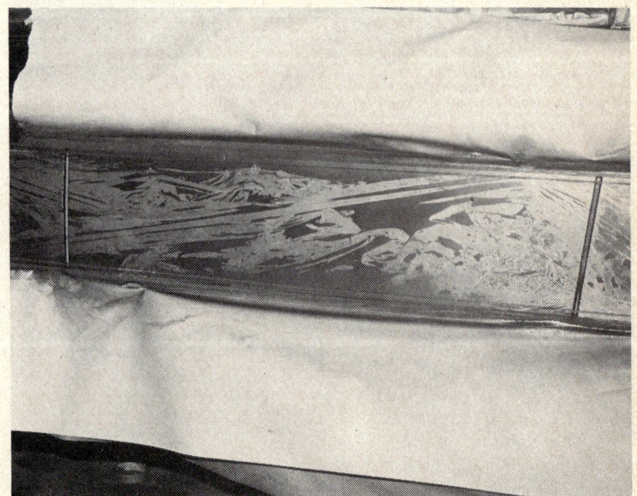

10 This time it looks like a surrealistic mountain scene in dark blues, light blues, metallic silver and blue-black. Should pattern not be pleasing, erase, respray and redo.

11 The finishing touch is obtained by carefully outlining both wood grain and mod mural panels with thinned black. Good control is required to reduce fogging.

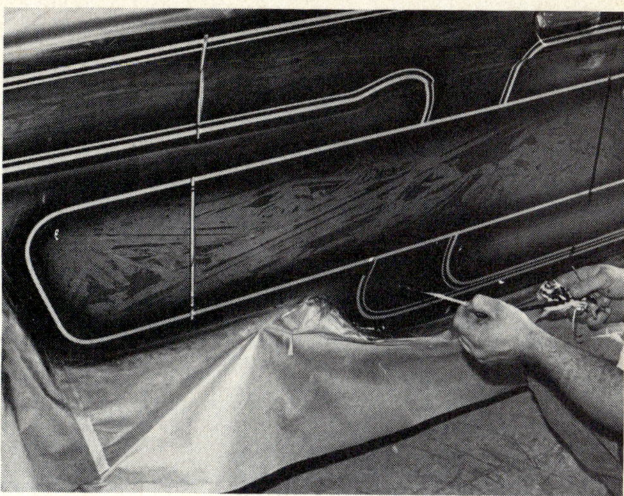

12 After allowing ample time for thorough drying, the ⅛-in. and ¼-in. masking tape is removed to reveal metallic silver and black pinstriping bordering the panels.

13 All areas are then sealed with a clear polyurethane (two-part) acrylic from Ditzler. This coating dries to a super-hard non-yellowing finish—the closest thing to epoxy.

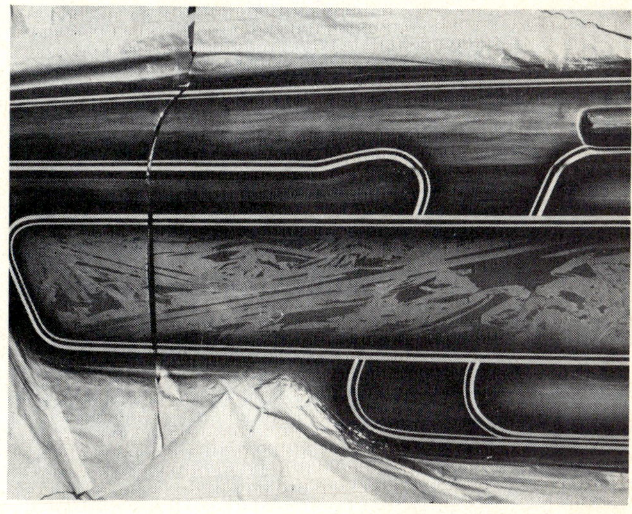

14 Needless to say, this clear coating has really punched up the colors in both the wood graining and the mod mural. The two-part polyurethane can be polished.

15 Here is the completed LUV truck with the borders still outlined by the masking tape and paper. Don't be too hasty in taking off the wrappings on the new present.

16 Oh alright, go ahead and unwrap it! Yep, it's another masterpiece from The Crazy Painters. For the full effect, turn now to page 175 and study photo No. 4 carefully.

HOW-TO: GOLD LEAF SHAPES

PHOTOS BY ERIC RICKMAN

Unlike pinstriping, which requires a special knack, gold leafing can be done by an amateur with satisfying results the very first time. And proficiency comes quickly with practice. Creativity? Well, that "does" require a special knack.

There are two kinds of gold leaf: real 23-karat gold leaf and imitation gold leaf. The real stuff is very, very thin and therefore more difficult to work with than imitation gold leaf. The real 23-karat gold leaf has a bright, beautiful appearance, but is quite expensive ($12 for a booklet of 3⅜-in.X3⅜-in. sheets). Imitation gold leaf doesn't quite match the lustre of the real stuff, but is considerably less expensive (a booklet of 5½-in.X5½-in. sheets for about $2). Other types of leaf available include copper, silver and variegated. This last item is by far the most popular with custom enthusiasts. It gives the appearance of gold that has been scorched with a torch (oil on water) and can be purchased with certain colors predominating in the pattern (blue, green, red or black variegated gold leaf).

The sizing is the most important ingredient in a gold leaf job next to the gold leaf itself. The sizing is a clear coating that dries tacky and becomes the adhesive that glues the leaf to the surface. There are two kinds of sizing: water base and oil base. The water base sizing dries faster than the oil base and is used when lacquer is to be applied over the finished job. Because the oil base sizing is an enamel, only enamels can be applied over the finished job.

Seelig's Custom Finishes, a mail order Mecca for custom painter's supplies sells, in addition to all types of gold leaf and sizing, a product called Varagold. This is peel-and-stick leaf (gold, silver, copper or variegated) with a clear plastic coating over it to keep it looking good for years and years.

1 Whether it's a body panel, motorcycle gas tank or a helmet, the surface must be completely clean. Nason Silicone Wax Remover is excellent for preparing the surface.

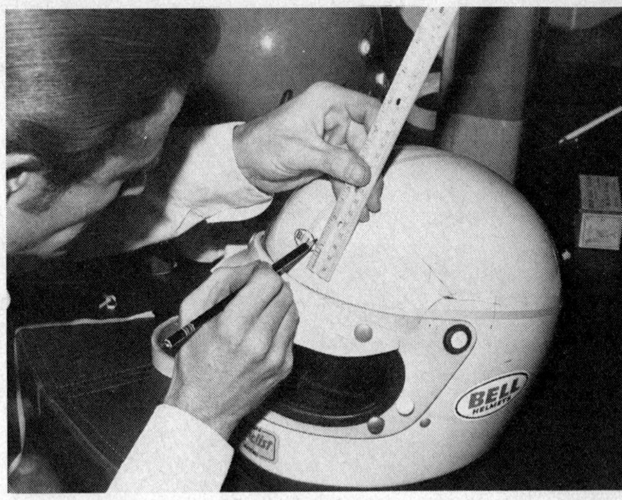

2 The easiest way to apply gold leaf to a geometric pattern is to mask off the outline. To guarantee symmetry, measure and mark the horizontal and vertical centerlines.

3 These centerlines are marked with a Stabilo pencil and connected with ¼-in. wide masking tape. This divides the work into uniform, easy-to-work-on, sections.

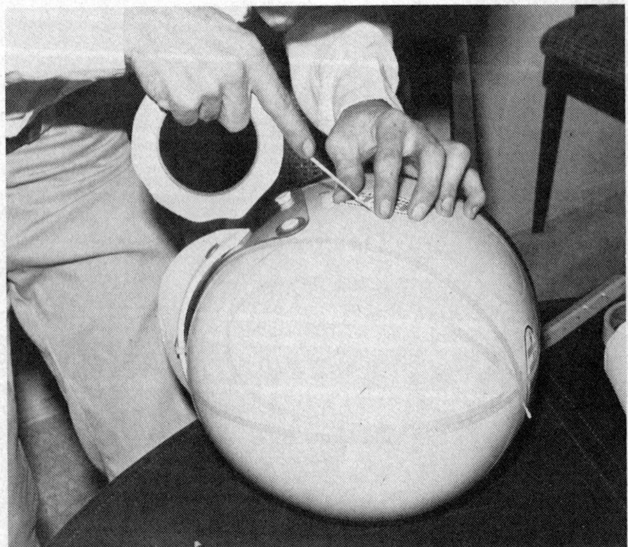

4 The design outline is next traced with ⅛-in. or ¼-in. masking tape. There's no need to cover areas to be left bare since sizing is painted on with soft camel hair brush.

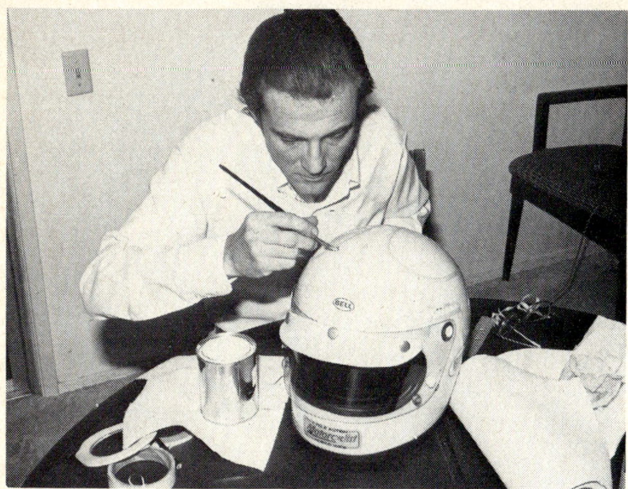

5 Wunda (water base) sizing is brushed on the area to be gold leafed. This sizing is milky looking but dries transparent in about 15 minutes. It dries completely in 6 hours.

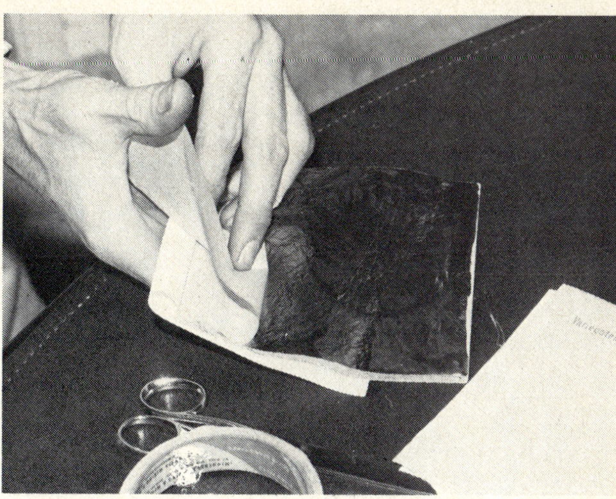

6 This job will be done with variegated imitation gold leaf and the next helmet with real 23-karat gold leaf. Each sheet has protective paper cover that is peeled back.

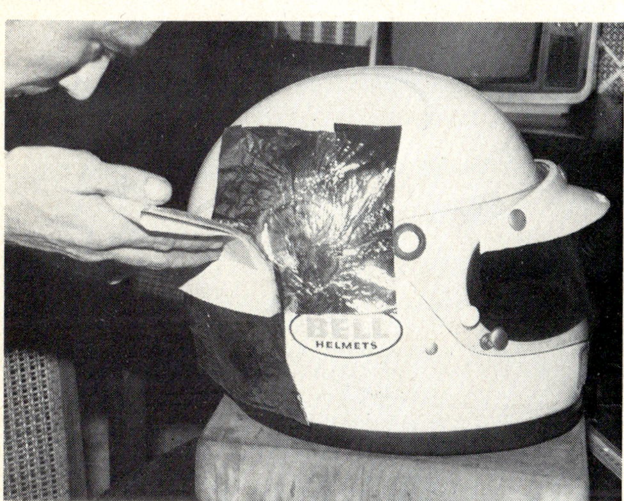

7 The whole booklet with exposed sheet of leaf up is laid against tacky sizing and the sheet of leaf sticks right on. The peel back protective paper and stick is repeated again.

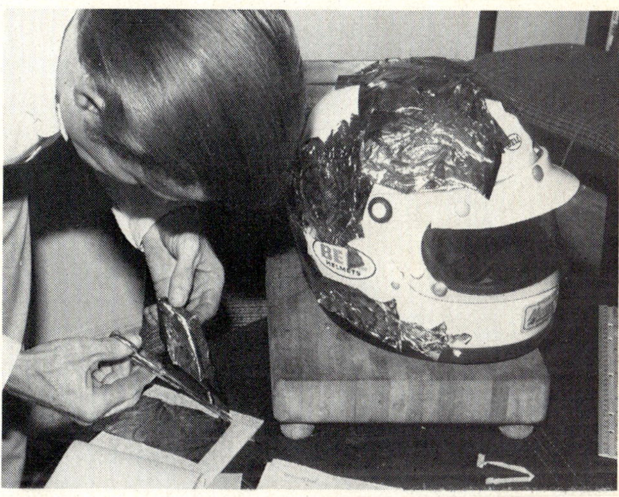

8 Sheets of imitation gold leaf can be cut carefully with scissors to handle small area, but don't try this with real 23-karat gold leaf—it's much too fragile to handle.

9 Sheets are applied in overlapping pattern—don't worry, no seams will be visible. Use large wad of cotton to gently pat variegated gold leaf to assure contact with sizing.

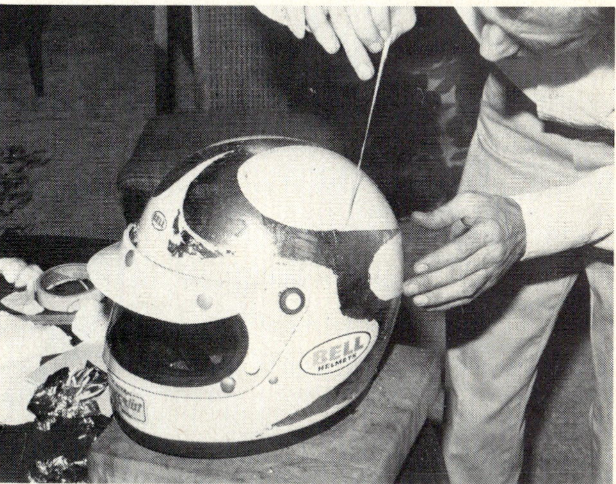

10 To reveal nice, crisp design the outline tape is merely pulled away. Clear lacquer coating will protect finish. Turn to page 175 (photo 9) for a look at leaf job in color.

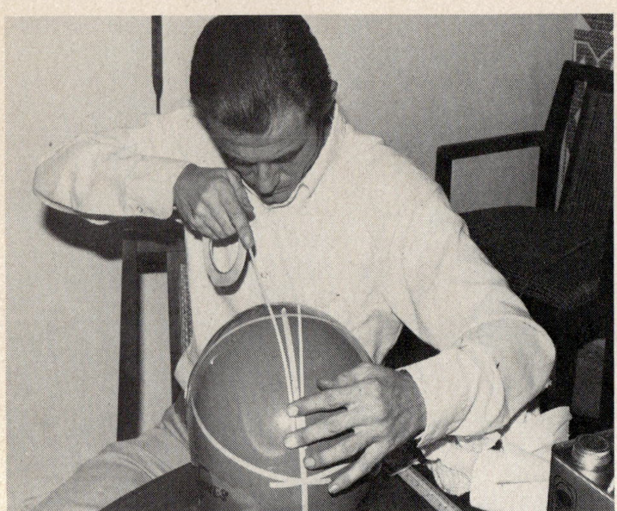

11 Applying 23-karat gold leaf follows same preparation of surface with cleaner and masking of Stabilo penciled outline with ¼-in. wide tape as with imitation gold leafing.

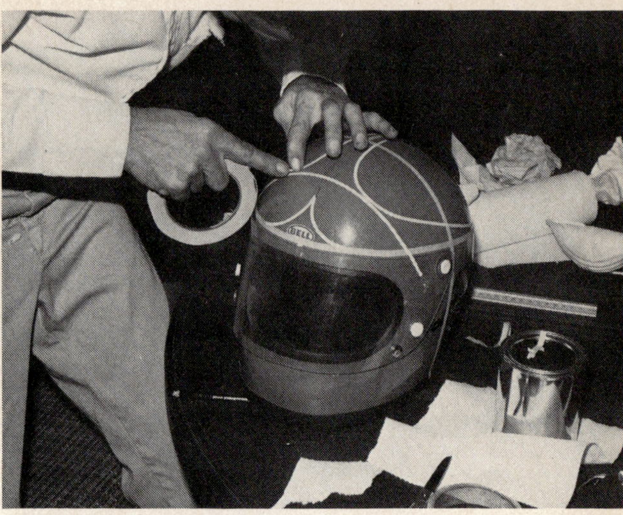

12 The big difference is use of oil base sizing to give more lustre to leaf. If tint is desired, use yellow ochre oil paint, not gold powder which may tarnish and darken.

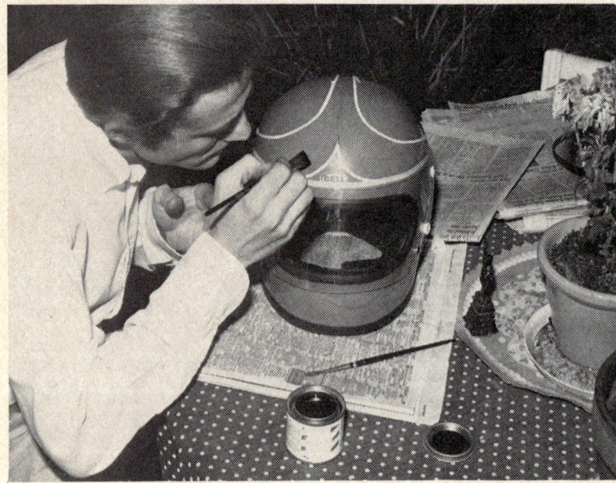

13 Completely cover area to be gold leafed with sizing and try to maintain a thin, even coating. A broad camel hair brush works well; Windsor-Newton sable is better.

14 A special wide brush called a gilder's tip is run over dry (not oily) hair and picks up the ultra-fragile sheet of gold leaf by static electricity and deposits it on size.

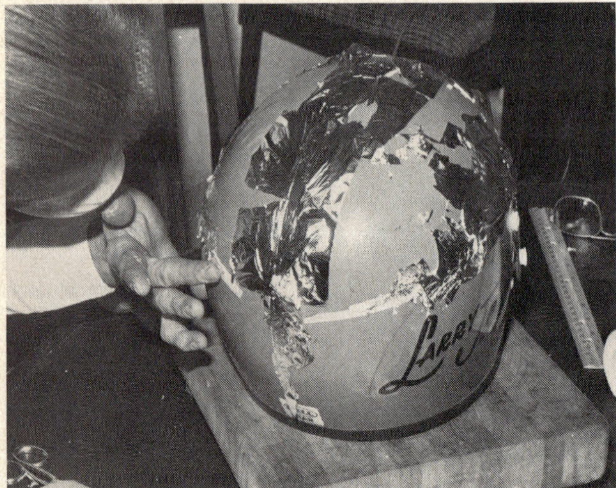

15 Oil base sizing becomes tacky in one hour, but wait until it levels out before applying leaf. Face mask is worn to avoid blowing delicate leaf away with the breath.

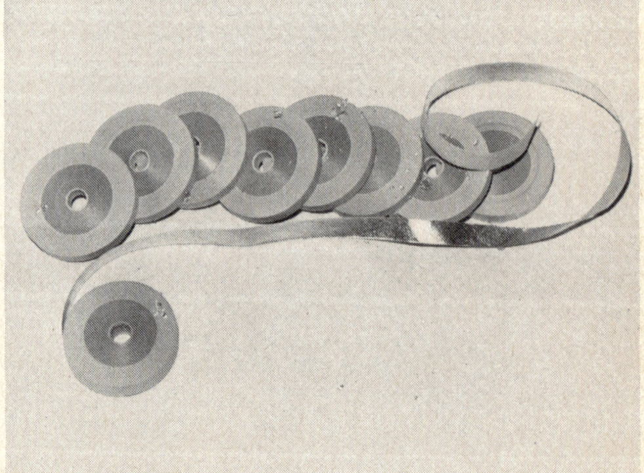

16 Seelig's Custom Finishes sells all manner of gold leaf supplies, including this 23-karat gold leaf tape. These and their gold leaf instruction booklet can be mail ordered.

HOW-TO: GOLD LEAF LETTERS

PHOTOS BY ERIC RICKMAN

Of course gold leafing can be applied to a freehand design and the outline taping is not really practical when lettering is involved. Many gold leafers (gilders!) trace the lettering on a sheet of paper (or design the thing freehand), then run over the outline with a pounce wheel—a toothed wheel that perforates the paper. Then the sheet of paper is positioned over the object to be lettered and patted with a pad of pounce powder. When the paper is removed, the design is transferred as a series of dots. Even very elaborate designs are easy to transfer using this technique.

The important part of the job is what happens after the gold leaf (real or imitation) has been applied. The leaf is affected by exposure to sun, rain, etc., so it must be sealed with a clear coating—lacquer, enamel, urethane, but not varnish. The customary approach is to outline the leaf with a pinstripe to bring out the gold leaf and seal the edges so moisture won't attack the sizing and lift the leaf. If some is good and more is better, then shadow outlining should be just enough—to really pop out that leaf job.

A trip to The Crazy Painters in Bellflower and a chat with Tom Kelly produced the enlightening step-by-step photos seen here and illustrates a professional's approach (along with some of his secrets) to a shadow-outlined variegated gold leaf logo. Ted Jones Ford in Buena Park, California, who quite often showcases (and sells) cars, vans and trucks individualized by The Crazy Painters, provided a beautiful black Ford Econoline so Kelly could demonstrate.

1 After cleaning off the wax and polish which could prevent the gold leaf from adhering properly, the location of the lettering is chosen and guide lines made with Stabilo pencil.

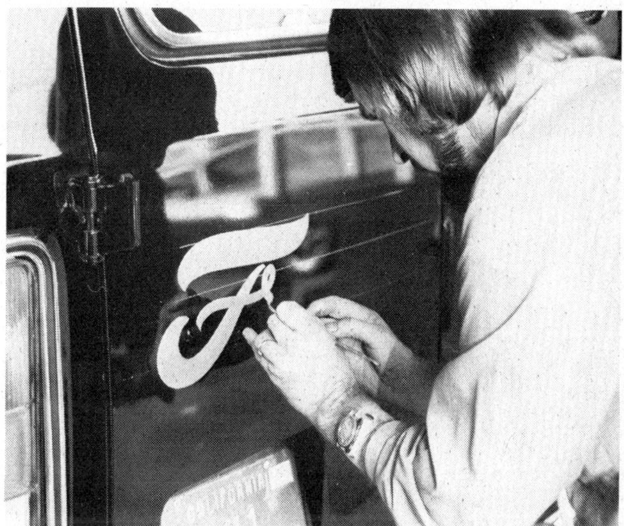

2 Gold powder is mixed with oil base sizing (tarnishing is no problem since variegated imitation gold leaf is used) and the flowing script is freehand painted on the van.

3 Guidelines in white Stabilo pencil can be painted on or over without harm. Skill as a sign painter, though, is what artist's finished job will look like when gold leafed.

4 Outline of each letter is done first, then brush loaded again with paint and center portion is filled in. Oil base sizing—like regular enamel—will level out after a while.

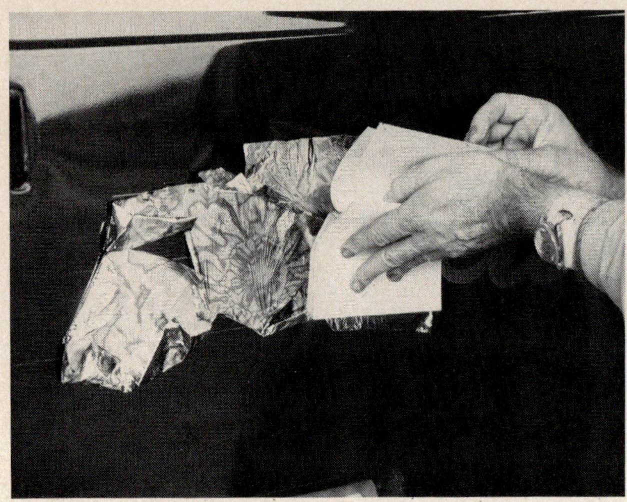

5 Sheets of leaf are loose in booklet so care must be taken not to let any drop out when peeling back cover paper and applying to sizing. Fingers push leaf against sizing.

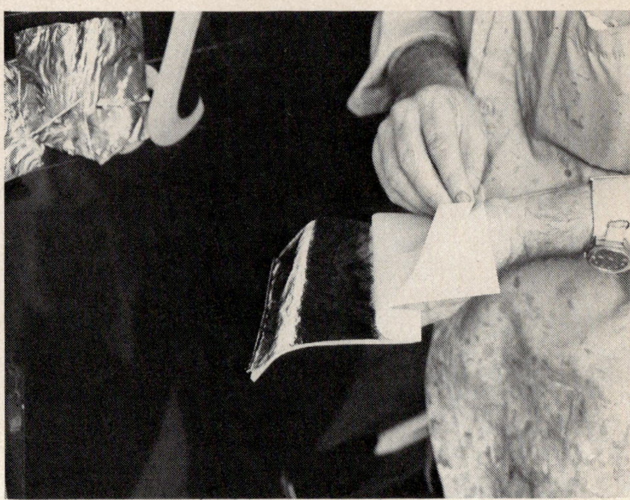

6 Thumb over top edge of booklet insures that sheets of variegated gold leaf don't drop out. If possible, center variegated pattern on the gold-tinted sizing for best look.

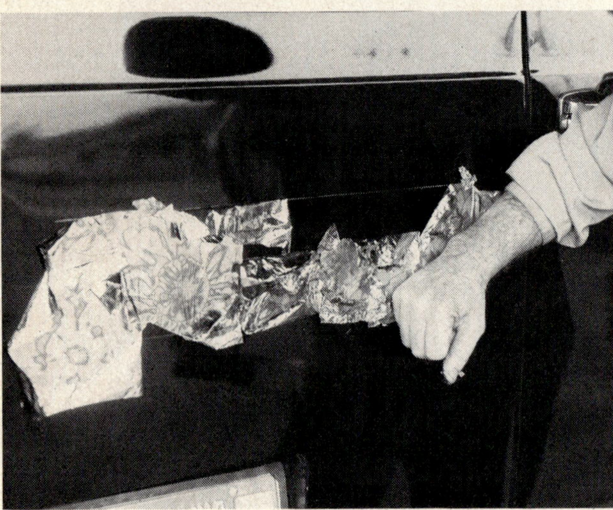

7 Side of hand is used to gently press sheets of gold leaf against the sizing. Be extremely careful not to twist hand or the fragile sheets will tear and wrinkle.

8 Rubbing with soft cloth (or soft thumb very gently) will remove excess leaf around edges of lettering. This excess can be used to fill in any uncovered areas if necessary.

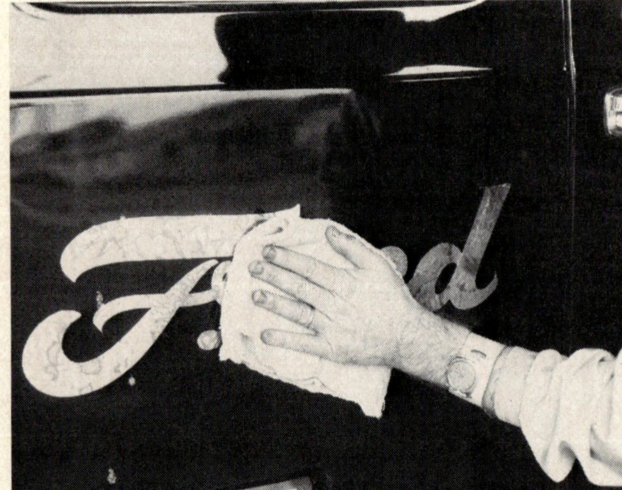

9 Burnishing gently with cloth will get rid of tatty little whisps of leaf clinging to edges and sharpen up the logo. Proper placement of sheets strengthens variegated effect.

10 Sealer (clear enamel) is immediately brushed on to bring out the beautiful coloration and to help seal leaf against effects of weather. Clear enamel is self-leveling, too.

11 Outlining of lettering on black van is done in yellow-orange One Shot Bulletin Enamel which is very resistant to weathering. Blue would also look nice with gold leafing.

12 Sign painter's rest (a stick with rubber tip) can be used for hand lettering by those who don't have nerves of steel, but our artist is an old master and steady, too.

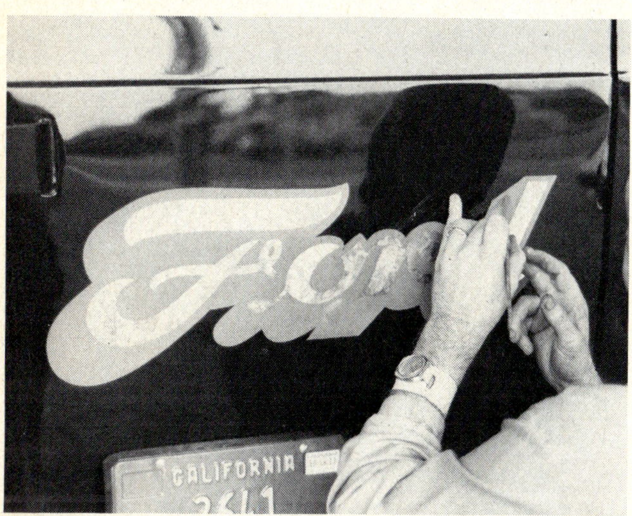

13 Shadowing is added next and first-class results require artist's sense of perspective. Imaginary light source is at upper right so letters cast uniform shadow to lower left.

14 Color is darkened a little and deeper shadows are added with Paasche VL airbrush. Bulletin Enamel is thinned 60/40 paint to thinner and pressure regulator set at 40 psi.

15 Yellow-orange outline is quite visible against shadows added with airbrush. Now white highlights are added to create 3-D look. Note reflection of airbrush in letter "r."

16 Completed masterpiece needs only one final touch: The artist's signature. Now we ask you to be the art critic and tell us whether shadow outlining resulted in work of art.

HOW-TO: GOP ART AND ACRYLUSTRATIONS

There are two processes by which a picture or any other art form printed on paper stock, can be lifted from the paper surface and the image then applied to a body panel much as a decal. Unlike a conventional decal though, you can apply virtually any illustration that strikes your fancy without being limited to only the subject selection that prepared decals offer. While the sky's the limit on selections made from magazine covers or pages, posters, brochures, or any other form in which either photos or hand art have been reproduced, you must be fairly selective in the quality of paper on which the illustration has been printed. Use only a good quality coated, glossy paper.

One of the processes for transferring a printed image to a body panel is known as GOP Art, its name coming from the use of Liquitex Gloss Polymer as the agent that softens and frees the image from the paper. The other technique is known as Acrylustration, a name possessed by Metalflake, Inc., since the process requires the use of their own brands of clear acrylic and lacquer thinner mixed at a 50/50 ratio.

If the illustration you wish to transfer is small—or if it's a large montage but made up of your own assortment of small pictures—don't cut out the desired image around its perimeter. Use either the entire page if it's normal magazine size (like this page), or trim a poster-sized piece of art down to about page size. Then, once the image area has been lifted, it can be trimmed to the desired outline and treated much like an ordinary decal from here on out.

Plan on spending some time in the whole process, especially when applying the transfer and working the bubbles or air pockets out from beneath it. The transfer is going to be a lot thinner than the ordinary decal, so while sqeegeeing the air out, you have to be very careful to not wrinkle the image. Watch.

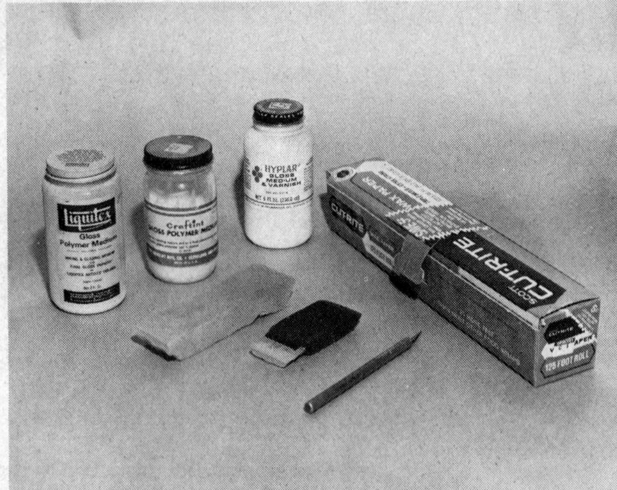

1 Here are the tools and supplies needed for the GOP Art transfer application; jars of Liquitex Gloss Polymer or Craftint Gloss Polymer, and Hyplar Gloss Medium and Varnish, an X-Acto knife, small squeegee and sponge, and wax paper.

2 Application of the Acrylustration-type transfer requires mixing at a 50/50 ratio Metalflake's house brands of lacquer thinner and clear acrylic. An ordinary spray gun can be used to apply this or the GOP Art mixtures, or an air brush.

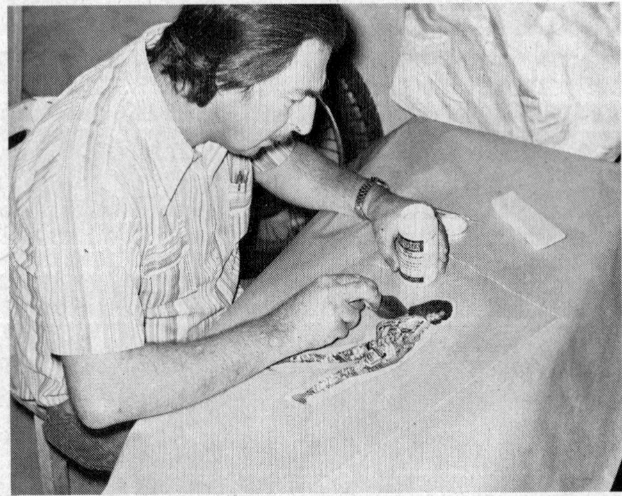

3 The desired image has been cut from a magazine and placed on a sheet of wax paper. For GOP Art, brush or sponge six coats (in alternate directions) of the polymer mixture, about 25 minutes between applications.

4 In Metalflake's Acrylustration method, the mixture of the two clears is lightly sprayed over the image. In the case of this small piece of art, the cut-out image wanted to blow away from air pressure. Entire page should be used.

5 After the GOP Art image has been allowed to dry overnight after the last polymer application, soak the transfer in lukewarm water until the paper can be peeled away from the image.

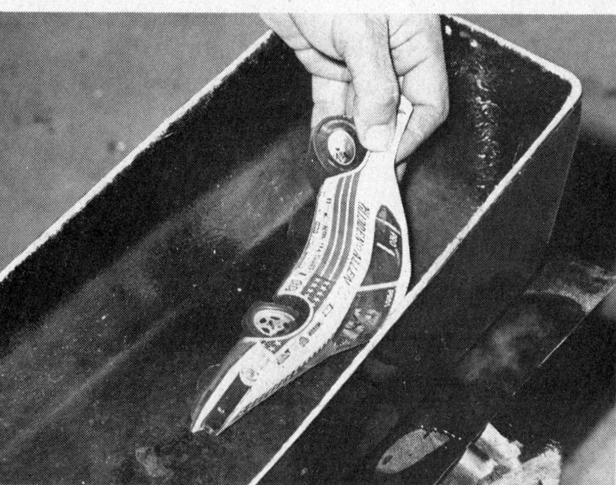

6 Acrylustration transfer requires the same step; soak the image in lukewarm water until the paper underneath the image can be peeled carefully away. The thicker (or more dense) the paper, the longer the soaking required.

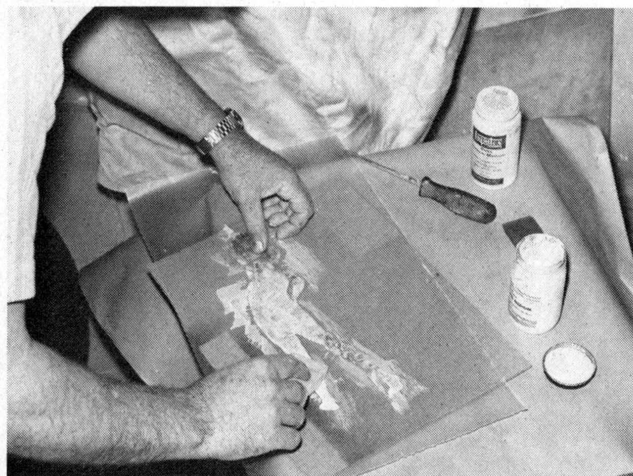

7 For GOP Art, the image must be allowed to dry over a second night (we said it takes time!) sandwiched between two sheets of wax paper, then the backside of the image is treated with another coating of polymer—but don't let it dry.

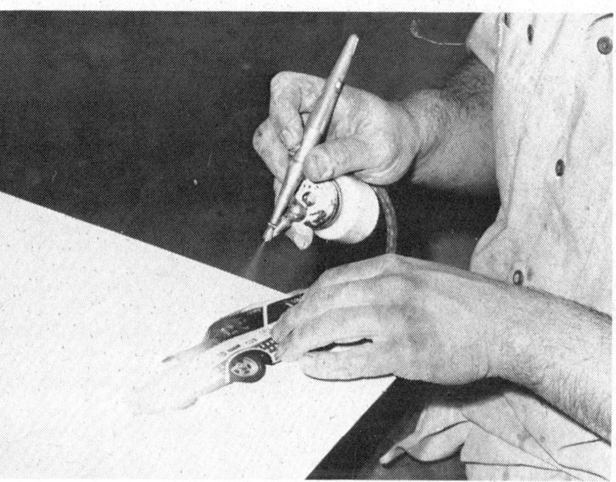

8 The image selected for the Acrylustration process proved to be from a low-quality printing job, and when the paper was peeled from it, the white tones were almost transparent. This was cured by air brushing white paint on the backside.

9 GOP Art image has been placed carefully on the surface, squeegeed to remove air and excess polymer from under it, then patted with cloth. Tree was air brushed on later by shooting through a stencil. Clear is now sprayed.

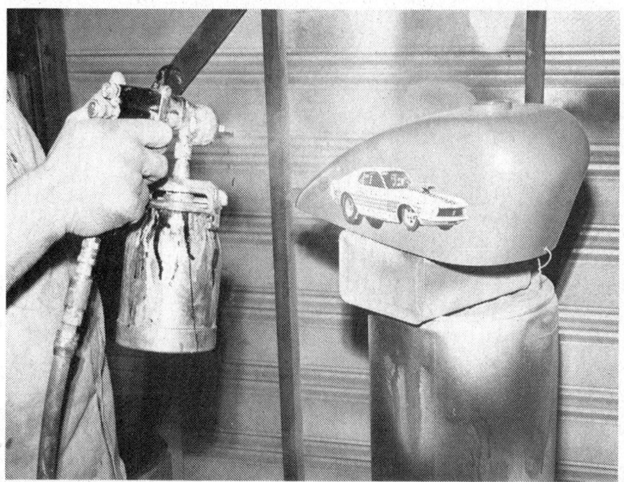

10 Acrylustration image is first sprayed on backside with a coat of the Metalflake mixture, applied to the surface and squeegeed down, then several more coats of clear sprayed over the image and surrounding area to seal edges.

PHOTOS BY ERIC RICKMAN

Styles come and styles go. Some people think that flame paint jobs went out with crew cuts and saddle shoes, but they are always much in demand.

To take a good look at this famous customizing technique, we visited The Crazy Painters in Bellflower, California. The vehicle to which the flames were applied was a little Datsun pickup truck from nearby Downey Datsun. The painting was done by Steve Young, while the final pinstriping was done by Bill Stowers, another Crazy Painter.

Before going into the actual sequential steps of flame painting, several important points must be mentioned. The first is that the area to be painted must be scrupulously clean. Steve wipes the area clean with thinner before sketching the flame outlines on the body.

Second, the flame areas must be lightly but thoroughly sanded. This must be done after masking, for if it is done before masking you will end up sanding areas which will not be repainted. These flame areas must therefore be dry sanded, as wet sanding will wet the masking tape and paper. This scuff sanding calls for a deft touch; it must be done right up to the edges of the masking, so that the flame paint will adhere, yet it must not be done so roughly that the crisp edge of the masking tape is destroyed.

This flame job was done in acrylic enamel, but of course lacquer can be used. Enamel was chosen because it was compatible with the paint already on the Datsun, and did not require a sealer or primer.

These flames were shot in varying shades of blue, starting with a light blue basecoat and progressively changing to a dark purple at the front of the hood and the side flames (color photo on p.172). Each successive darker tone was carried to a point just short of the previous one, and the flame tips were tinged with orange. The finished job was then covered with several coats of acrylic clear to which a small amount of red-violet pearl was added for sparkle.

1 After carefully cleaning the approximate flame area with thinner, Steve Young of The Crazy Painters sketches out the flames with a china marker pencil. Stay away from all chrome and trim pieces, so you don't have to remove them.

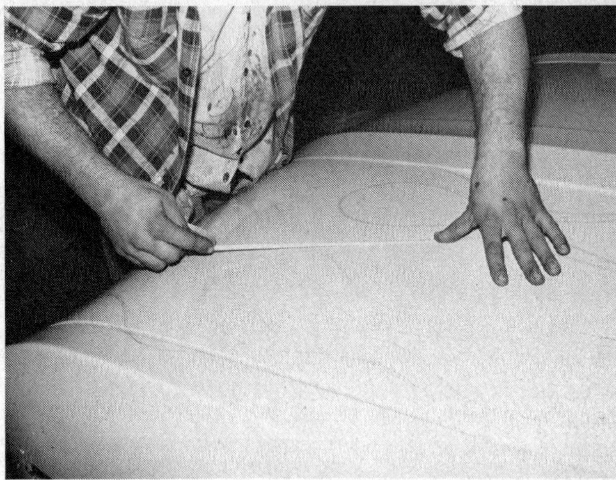

2 Once the pattern has been decided, Steve then begins the masking by using ¼-in. masking tape to follow the curved outlines of the flames. This thin tape is flexible enough to follow the twisting flames without lifting or splitting.

3 Following the ¼-in. tape with a single layer of ¾-in. tape, Steve then seals the tape with a small rubber roller. Such rollers are available from photo supply houses, and are called "ferrotype rollers" since they squeegee ferrotype plates.

4 With the flame outlines taken care of, the small non-flame areas near the flames are then masked with normal ¾-in. tape. This could just as well be masked with paper, but Steve finds it faster to just use lots of tape.

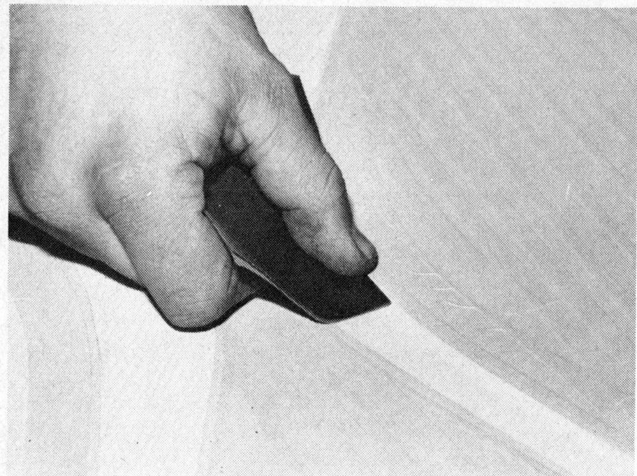

5 Emery cloth or wet-and-dry sandpaper should be folded over double, and the edge used to carefully sand the sides, edges, and tips of the flame areas. Be careful here, as carelessness will ruin the sharp crisp edge of the tape.

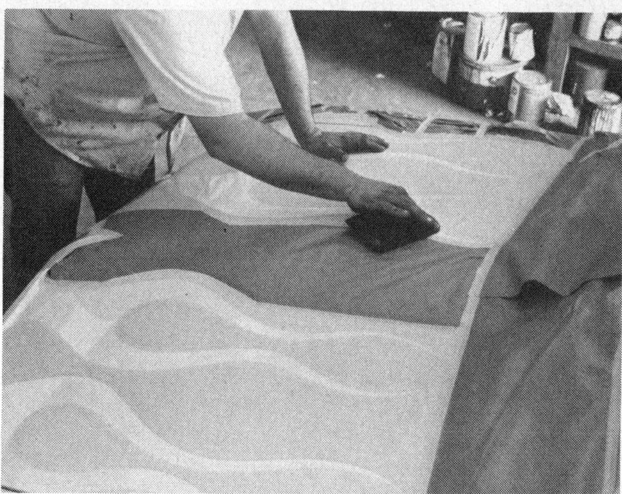

6 With all tape applied, the rest of the vehicle is masked in the normal manner with tape and masking paper. After sanding an area, Steve is careful not to rest his hand on it, as the oils in the skin may cause improper paint adhesion.

7 The light blue basecoat is now sprayed over the entire flamed area. Since this was acrylic enamel, the drying time between coats was rather lengthy. Mixing in a catalyst or using acrylic lacquer considerably speeds the process.

8 The complete outline of the side flames can be seen in this view. Note how the flowing shape follows the front wheelwell, then artfully avoids all trim pieces. Steve must have used up two rolls of tape on this job.

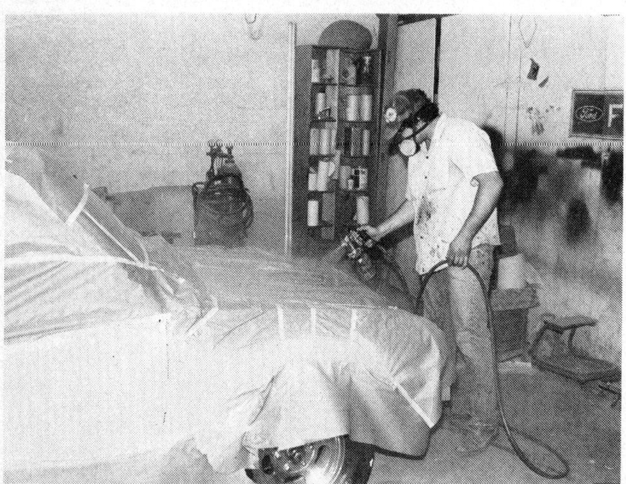

9 Carefully limiting the area he covers, Steve sprays over the blue base with progressively darker shades. The secret of a really good progressive color change is to carry each darker color right down to the base, blending in at the top.

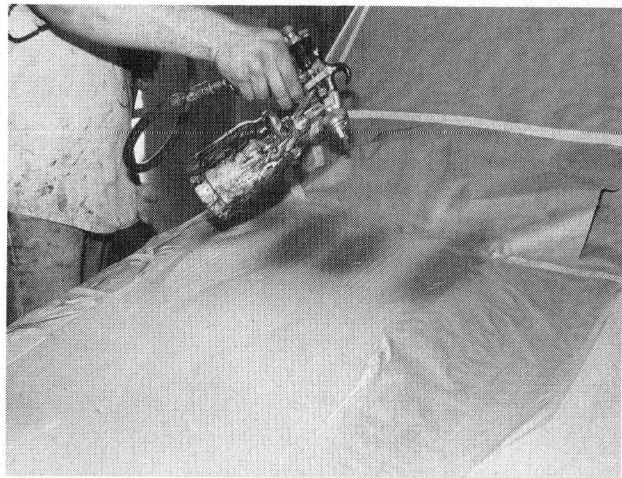

10 It doesn't look quite right in this black-and-white photo, but that is orange being sprayed on the flame tips. When all color was dry, a topcoat of clear mixed with red-violet pearl was sprayed for fading protection and sparkle.

11 The moment of truth arrives as the masking tape is finally removed. With enamel, it should be removed immediately after shooting, so edges will "heal" themselves. Note original sketch lines left on the car during painting.

12 Certainly the most exciting moment for painter and owner is when the masking is pulled away and the total effect of the customizing can be seen. Steve's done this enough, so he merely casts a critical eye on his handiwork.

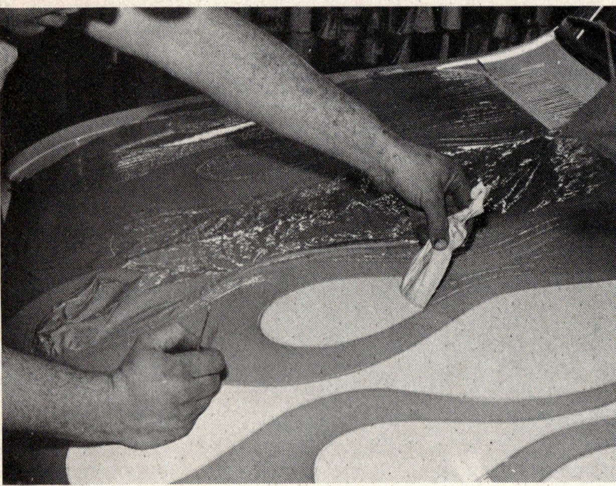

13 It may have been a little more trouble to use all that tape in masking around the flames, but it's a lot easier to remove all at once than a combination of tape and masking paper. Note glossy flames from pearl topcoat.

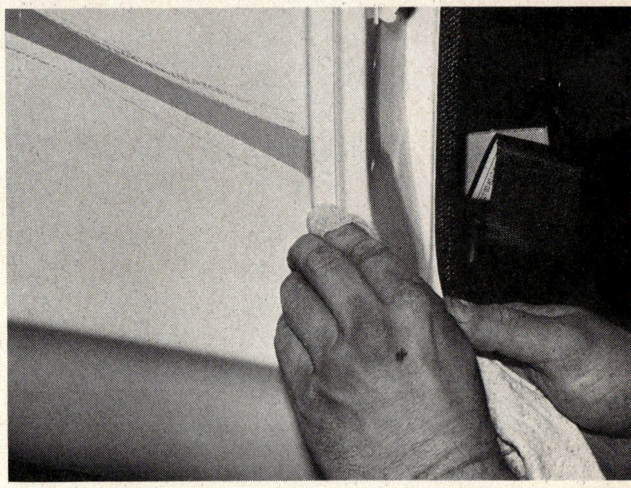

14 With masking removed from the whole car, Steve goes over the Datsun with a rag soaked in thinner to remove the china marker sketch lines. There's no danger of smearing, as dry acrylic enamel is impervious to thinner.

15 And here's the total effect, straight out of the Fifties. Flames can be any color, but the reason that Crazy Painters chose a progressive scheme was that a solid-color flame would have lost effect on truck's long side.

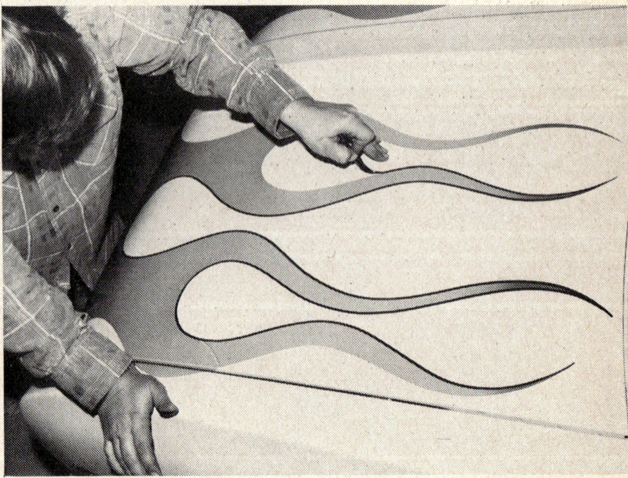

16 Any flame job should be finished off by striping the flame outlines, in this case with black. Striping heightens the dancing effect of the flames, and also serves to cover the raised edge of the flame and blend it into the body color.

Designer's Showcase

The difference in painting is the difference in painting.

by Al Hall

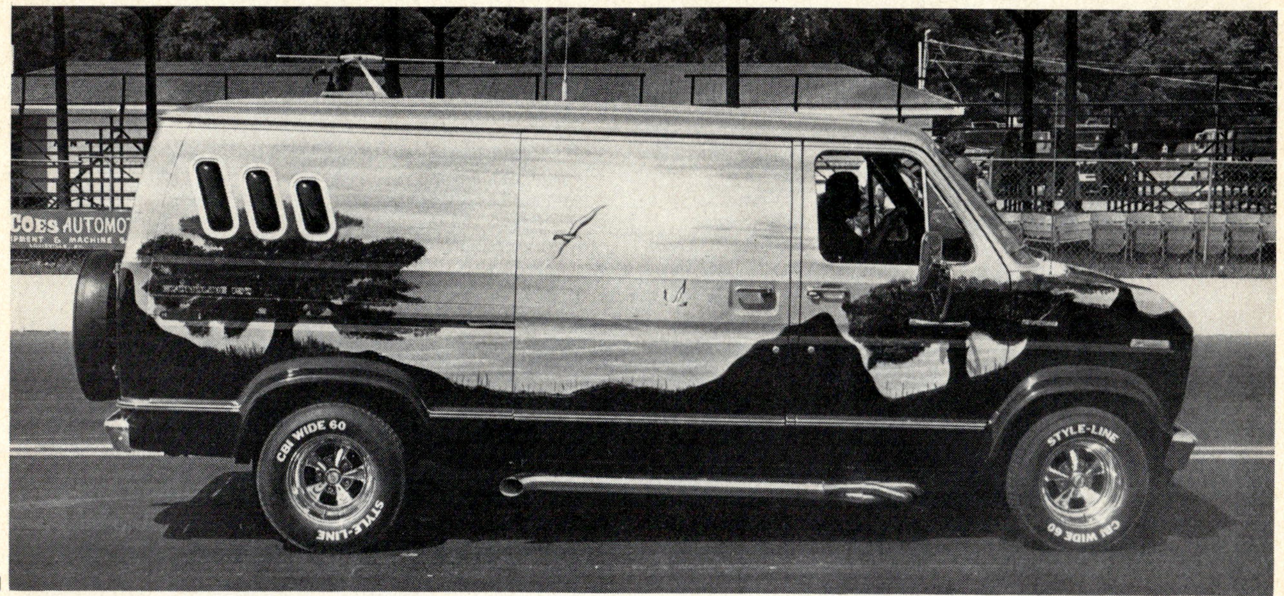

1

2

3

1. Every facet of the mural-painting genre seems to have been covered by now, from small panels on the upper quarters to full designs with science fiction themes, etc. This treatment with the mural going around the whole van and the base color functioning as the background is very popular now.

2. Wilder designs certainly attract a lot of attention, but when you have to look at your own vehicle every day, the taseful approach, such as this interesting treatment on an early van, is a comfortable one to live with.

3. Here's some unusual paint work! This pickup features a center side panel filled with psychedelic designs sprayed over various templates, and surrounded by a pastoral mural scene cut up into short "strobe" sections.

4. Called the Tijuana Taxi, this wild gull-wing '50 Chevy was built from a wrecked four-door. Color Me Kustom applied a base of silver flake and topped this off with wild strips of candy reds, "marbling" and dabs of silver over black panels and a mural.

4

Designer's Showcase

1

2

3

4

5

1. This basically white van gets its stares thanks to an array of shaded candy "spurts" of color. Different!

2. Here's a technique we've never seen before, but one that has great possibilities. This huge mural of a watchful lion seems held back only by "bars" of the stock van paint. Your first look at this van in person really gets your attention quickly.

3. Car Craft magazine's Pacer project was finished off with a "graphics" style two-tone job of yellow and maroon.

4. Sponsor-conscious racers are now turning to clean-but-striking paint treatments for their race cars and the street machiners are picking up on the idea, too. This chopped Chevy sedan delivery has graphics everywhere.

5. Vans customized on a production line for dealerships usually exhibit bold, angular scallops or panelling that has become really tiresome. This custom van however, treats the lines differently by having them weave in and out of each other, and adding a mini-mural on the side doors.

6. Seen here at a van meet, this E-100 Ford is about as graphic as you can get. Only the roof is left un-muralled. Bobbie Beam did the Wizard of Oz work in soft pastel colors outlined with sharp black lines over a yellow pearl base. This striking job cost $3,000!

7. Flames have been around so long, you'd think there would be no more new ways to utilize them, but here's a new look on this stepside pickup.

8. Stars and stripes and American flags have always been popular for vans and Corvettes, but this AMX street machine turns Old Glory into a new way to look at flames.

6

7

8

Creative Color

Showmanship

Custom bodywork deserves a custom paint job, but custom painting alone can transform an ordinary stocker into a highly individual piece of automotive art. Like the bodyman, the custom painter requires artistic ability as well as technical skill. **1)** In the Fifties, talented young artists such as Donn Varner from Eugene, Oregon, turned stockers into customs and customs into showpieces with imaginative pinstriping. Every area of the country became fertile ground for those who had mastered the almost forgotten technique of hand-painting lines on car bodies. **2)** In San Diego, Ray Cook's four-color flame design on Bob McCoy's '40 Ford Tudor created a sensation and was widely copied. Ray softened the color transitions with rubbing compound and outlined the pattern with white pinstriping. **3)** In addition to scallops, stripes and Oldsmobile "Fiesta" hubcaps, this Chevrolet Pickup boasted a surrealistic mural on the tailgate long before custom vans became a canvas for frustrated Frazettas. **4)** Car shows mirror the trends in custom painting as here at the 1964 Oakland Roadster Show. Cars belonging to members of East Bay Rods run the spectrum from Metalflake, to candy, from flames to paneling. **5)** Elaine Sterling and Larry Watson pose with Larry's customized '50 Chevrolet Fleetline which was painted by Damon Richey in custom mixed purple lacquer called "Grapevine." Watson gained lasting fame later as a pinstriper and custom painter. **6)** Lucky is the owner of a Von Dutch original on canvas! Sale of such paintings was not unusual at custom car shows in the Fifties, and individual murals on Tee-shirts were even more popular, but who today can boast of having an original Ed "Big Daddy" Roth caricature Tee-shirt? **7)** Whether it is new or old, custom or stock, the paint job is the finishing touch and the first thing to catch the onlooker's attention.

1

SPENCE MURRAY PHOTO (1956)

2

BOB HARDEE PHOTO (1956)

3

4 PPC PHOTOGRAPHIC PHOTO (1964)

5 BOB D'OLIVO PHOTO (1956)

6 AL HALL PHOTO (1956)

7 RALPH POOLE PHOTO (1961)

Creative Color

Basic Beauty

Forget those axioms about painting small cars in bright colors and large cars in darker hues, or using lighter shades if the body panels aren't dead smooth. Custom colors—even light ones—magnify poor bodywork. Thought and thorough preparation are the prime ingredients for a unique custom finish. **1)** Eddie McCullough's '75 Ford van displays a beautiful Candyapple lacquer finish with variegated gold leaf striping and sign. **2)** Speaking of thought and preparation, imagine the amount of both that were lavished on the 'T' by owner/painter Jerry Lee Morris and his father. Wood graining utilizes powder technique described in previous chapter. **3)** Gene Winfield uses his gold-plated Binks to spray sealer over Aero-Lac candy blue. **4)** Aero-Lac Design Colors can be used to achieve beautifully unusual effects for panels or borders. **5)** A.J. Foyt's 1961 Indy entry was sprayed in pearl white with Candyapple nose by Dean Jeffries, who's seen at left adding finishing touches. **6)** Otto Rhodes and Marian Mesojedic pose with Otto's 1956 Ford pickup finished in Iridescent Pearl with Metalflake red pickup box by G&D Body & Paint of Colorado. Pearl is excellent choice for any vehicle with abundance of compound curved surfaces. **7)** Vinnie Bergeman's Vandemonium is a symphony in Metalflake and candy colors. R&S Color in Anaheim spent 75 man-hours on the candy over silver Metalflake mural alone. From the substrate to the final clear topcoat, there are 100 coats of paint and much block sanding between coats.

1 BOB McCLURG PHOTO (1977)

2 ERIC RICKMAN PHOTO (1961)

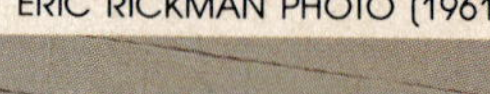

3 JAY STORER PHOTO (1978)

4 JAY STORER PHOTO (1978)

5 ERIC RICKMAN PHOTO (1961)

6
TONY SPICOLA PHOTO (1962)

7
BOB McCLURG PHOTO (1976)

Creative Color

Stripes & Flames

Custom painting need not be done from the substrate (metal or fiberglass) on up. Pinstripes and flame paint jobs can be applied quite effectively to a factory finish in good condition. And the effects can vary anywhere from mild to wild depending on the owner or painter's whims. **1)** Steve Young of The Crazy Painters in Bellflower, California, turned this stock Datsun King Cab into something neat for the street with a bit of imagination and acrylic enamel. **2)** This five-color flame job was sprayed one color over the other by Damon Richey, who then rubbed out successive colors on a selective basis to achieve this unusual effect. **3)** Modern application of traditional flame job by Larry Watson on Bill Branch's 1946 Ford was detailed step by step in Hot Rod Magazine's Rod & Custom No. 3. **4)** This 1967 Oakland Roadster show entrant went the flame route in Metalflake and candy colors. **5)** Bold pinstriping and gold leaf was quite popular in late Fifties for lifting car out of stock status. **6)** At today's prices, there's a million dollars worth of metal on this high school football field. Two-color flames make roadster a standout. **7)** Mildly customized 1956 Chevrolet is wildly pinstriped in two colors. **8)** The most outrageous flame job ever done was applied to Earl Bruce's Mercedes-Benz 300-SL Gullwing coupe by none-other than Von Dutch. Note exquisite pinstriping and multicolor tips on flames. Some viewers said the car looked like a hot fudge sundae on wheels.

ERIC RICKMAN PHOTO (1977)

2

AL HALL PHOTO (1958)

3

GREG SHARP PHOTO (1977)

4 ERIC RICKMAN PHOTO (1967)

5 AL HALL PHOTO (1957)

6 AL HALL PHOTO (1957)

7 AL HALL PHOTO (1956)

8 AL HALL PHOTO (1956)

Creative Color

Scallops, Et Cetera

For the most part, custom car enthusiasts avoided flame paint jobs. Their preference tended toward scallops, paneling, fading and shading. From this beginning has evolved most of today's painting techniques. **1)** Candy panel shadowing over white-gold Murano pearl is the handiwork of James Hurley and his La Habra high school body and paint shop students. Precise overlap of orange, blue, purple and green candy colors creates additional hues where patterns cross. **2)** Modern tastes tend toward bold blocks of color as on this Datsun 240-Z sprayed in candy over pearl. **3)** This early Sixties style paneling has been revived by vanners. **4)** Steve Young (The Crazy Painters) combined wood graining and Saran-Wrap muraling on this stock Chevrolet Luv. How-to is detailed in previous chapter. **5)** Custom Corvette Wagon combines cinnamon brown with salt and pepper gold Metalflake and black outlining. Paint is by owner Jim Ash of 'Vette Shop in Anchorage. **6)** Unique panel painting in Larry Lee Dere's 1969 Camaro was applied by Mike Lowery of Anchorage. **7)** Brent Morgan of the Fantasy Factory in North Hollywood shoots strong candy gold over silver, fogs on black lacquer with a bit of enamel reducer added and wipes this with lint-free rag for wood-grain effect. This is followed by clear with a bit of gold toner in it. **8)** The Bob Eldridge technique of wood graining with powders (detailed in previous chapter) gives realistic results. **9)** Gold leafing is best confined to smaller areas, but is effective in combination with patterns or alone.

1

ERIC RICKMAN PHOTO (1977)

2

JAY STORER PHOTO (1977)

3

4 ERIC RICKMAN PHOTO (1977)

5 RICK RICKETSON PHOTO (1977)

6 RICK RICKETSON PHOTO (1977)

7 ERIC RICKMAN PHOTO (1977)

8 ERIC RICKMAN PHOTO (1977)

9 ERIC RICKMAN PHOTO (1977)

Creative Color

Mural Magic

Von Dutch painted a surrealistic mural on the side of Barris Kustoms' 1954 Ford parts wagon over 22 years ago, and many's the pickup tailgate that has been likewise adorned in the intervening years. However, vanners definitely popularized the ideas and has it spread. **1)** Final look at the Kelly mural on our how-to van kindly provided by Ted Jones Ford in Buena Park. **2)** Intricate ideas require plenty of pre-planning if the final result is to tease the eye. **3)** Mr. Bottjen of Keystone, S.D., accepted the challenge posed by Jim Harris from Richardson, Alaska to do a winter mural on his white Dodge Van. **4)** The Crazy Painters in Bellflower gave the owner of this Pontiac Firebird a custom painted Phoenix and pinstriping too for little more than the cost of a decal alone. **5)** Another example of imagination from Jim Ash's 'Vette Shop in Anchorage, Alaska, is this desert mural on Holly Anderson's 1966 Corvette. Pearl yellow lacquer sprayed by Ash was striped in gold and orange.

1

ERIC RICKMAN PHOTO (1977)

2

MIKE PARRIS PHOTO (1964)

3

RICK RICKETSON PHOTO (1977)

4

ERIC RICKMAN PHOTO (1977)

5

RICK RICKETSON PHOTO (1977)